Post Harvest Technology & Processing of Horticulture Crops

AF648869

NIPA® GENX ELECTRONIC RESOURCES & SOLUTIONS P. LTD.
New Delhi-110 034

Post Harvest Technology & Processing of Horticulture Crops
At a Glance

[Includes Previous Year Questions of JRF (ICAR) and Ph.D. (IARI)]

P.S. Pandit

Bhani Ram

Tanveer Ahmad

Centre of Excellence on Post Harvest Technology
Navsari Agricultural University
Eru Char Rasta, Navsari
Gujarat - 396450

NIPA® GENX ELECTRONIC RESOURCES & SOLUTIONS P. LTD.
New Delhi-110 034

NIPA® GENX ELECTRONIC
RESOURCES & SOLUTIONS P. LTD.

101,103, Vikas Surya Plaza, CU Block
L.S.C. Market, Pitam Pura, New Delhi-110 034
Ph : +91 11 27341616, 27341717, 27341718
E-mail: newindiapublishingagency@gmail.com
web: www.nipabooks.com

For customer assistance, please contact
Phone: + 91-11-27 34 17 17
Fax: + 91-11- 27 34 16 16
E-Mail: feedbacks@nipabooks.com

ISBN: 978-81-19254-67-5

Composed and Designed by NIPA®.

Dedicated to

Post Harvest Technologists,
Students and Scientists

Preface

India is second leading producer of fruits and vegetables but after that less than 2% of fruits and vegetables utilized in processing, it is very less as compared to developed countries, post harvest losses are very high, that is (20-40%) due to lack of proper post harvest handling. But other side of this demerit is creating a huge scope in this field not only for people those are relevant but also to others.

Contemporary time were the times of campaign, but now a days competitive examinations have became regular feature and also more and more people and organisations are becoming aware about post harvest technology and its fruitful future. So as a consequence there are a large number of aspirants who have wished to adorn their career through post harvest technology. The present book "Post Harvest Technology & Processing of Horticulture Crops : At a Glance" cover almost all valuable chapters regarding subject matter with Multiple Choice Questions and Queries. The attempt has been made to deliver a lot of subject's information to compete at various examination like ARS, NET, JRF, SRF, Ph.D., FCI as well as competition of UPSC and State PCSs for various jobs.

The authors consider that the readers will get ample information concerned with the post harvest technology and preservation of fruits, vegetables, flowers, spices and aromatic crops. There may, however, be still some shortcomings in this book and authors will be really grateful to receive proposal from readers for incorporation in the next edition of this volume.

P.S. Pandit

Bhani Ram

Tanveer Ahmad

Acknowledgement

The authors are indebted to Dr. A. R. Pathak, Hon. Vice-chancellor, Navsari Agricultural University, Navsari and Dr. A. N. Sabalpara, Director of Research and Dean P.G. Studies for inspiring to write a comprehensive book for graduate and post-graduates, who are preparing for competitive and various entrance examinations like; NET, ARS, JRF, GRE, SRF etc. The authors acknowledge the thanks of gratitude to Er. S. P. Shukla, Incharge, Centre of Excellence on Post Harvest Technology, Navsari Agricultural University for suggesting the name of M/s. New India Publishing Agency, New Delhi as publisher of the book. The authors are grateful to Mr. Durga Singh, Mr. Dayanand Yadav, Mr. Arvind Chita and Mr. Tarachand Simi for helping in typing the manuscript of the book.

Parag Pandit
Bhani Ram
Tanveer Ahmad

Abbreviation

a_w	:	Water Activity
ACC	:	1-Aminocyclopropane-1-carboxylic acid
ACD	:	After Cooking Discoloration
AGMARK	:	Agricultural Produce (Grading and Marking) Act, (1937)
AOA	:	Amino Oxy Acetic Acid
APEDA	:	Agricultural and Processed Food Products Export Development authority
AR	:	Acid Resistant
ATP	:	Adenosine triphosphate
AVG	:	Aminoethoxy Vinyl Glycin
BA	:	Benzoic Acid
BHA	:	Butylated hydroxyanisol
BHT	:	Butylated hydroxytoluene
BIS	:	Bureau of Indian Standard
BOD	:	Biological Oxygen Demand
BVO	:	Brominated Vegetable Oil
CA	:	Controlled Atmospheric
CAC	:	Codex Alienators Commission
CACP	:	Commission for Cost and Price
CCC	:	Cycocyle (Chlorocholine Chloride)
CCFS	:	Central Committee for Food Standards
CCP	:	Critical Control Point
CFB	:	Corrugated Fiber Board
CFL	:	Central Food Laboratory (Kolkata)
CFR	:	Code for Federal Regulation (US)
CFTRI	:	Central Food Technological Research Institute, Mysore

CIPHET	:	Central Institute for Post Harvest Engineering and Technology, Ludhiana, Punjab
CMC	:	Carboxymethylcellulose
DHA	:	Dehydro Ascorbic Acid
DLF	:	Delayed Light Emission
DMI	:	Directorate of Marketing and Inspection
EDITA	:	Ethylene Diamine Tetra-acetic Acid
EMC	:	Equilibrium Moisture Content
ERH	:	Equilibrium Relative Humidity
EVA	:	Ethylene vinyl acetate
FAO	:	Food and Agriculture Organization
FDA	:	Food and Drug Administration
FHA	:	Food Health Authority
FPO	:	Fruit Product Order (1955)
GC	:	Gas Chromatography
GAP	:	Good Agriculture Practices
GDP	:	Gross Domestic Production
GHP	:	Good Hygiene Practices
GM Food	:	Good Modified Food
GMP	:	Good Manufacturing Practices
GRAS	:	Generally Recognized As Safe
HACCP	:	Hazard Analysis Critical Control Point
HB	:	Hydroxybutyrate
HDEP	:	High Density Polyethylene
HHP	:	High Hydrostatic Pressure
HMFP	:	High Moisture Food Products
HPC	:	Hydroxy Propyl Cellulose
HPLC	:	High Pressure Liquid Chromatography
HPMC	:	Hydroxy Propyl Methyl Cellulose
HV	:	Hydroxyvalerate
HTST	:	High Temperature Short Time
IFS	:	International Food Standard
IMF	:	Intermediate Moisture Food
IPC	:	Isopropyle carbamate

IQF	:	Individual Quick Freezing
ISO	:	International Organization of Standardization
ISP	:	Isolated Soyabean Protein
KMS	:	Potassium Meta Bisulfide
LOX	:	Lipoxygenase
LDEP	:	Low Density Polyethylene
LTHT	:	Low Temperature High Time
MAP	:	Modified Atmospheric Packaging
MAQ	:	Minimum Acceptable Quality
MC	:	Methylcellulose
MCT	:	Microbial Challenge Testing
MFPO	:	Meat and Food Product Order (1973)
MH	:	Maliec Hydrazide
MMPO	:	Meat and Meat Product Order (1972)
MPEDA	:	Meat and Marine Food Product Export Development Authority
MPF	:	Minimally Processed Food
MSG	:	Monosodium Glutamate
MTP	:	Multi Target Preservation
MVTR	:	Moisture-Vapor Transmittance Rate
NEB	:	Non Enzymatic Browning
NMR	:	Nuclear Magnetic Resonance
OD	:	Optical Density
PA	:	Polyamides
PE	:	Polyethylene
PEF	:	Pulsed Electric Field Technology
PET	:	Polyethylene Terephthalate
PFA	:	Prevention of Food Adulteration Act (1954)
PG	:	Poly Galactouronase
pH	:	Potential of Hydrogen
PHM	:	Post Harvest Management
PM	:	Predictive Microbiology
PP	:	Poly Propylene
PPF	:	Partially Processed Food

PPM	:	Part Per Million
PPO	:	Poly Phenol Oxide
PS	:	Polystyrene
Psi	:	Ponds per square inch
Pusa-ZECC	:	Pusa Zero Energy Cool Chamber
PVC	:	Poly Vinyl Chloride
QA	:	Quality Assurance
QC	:	Quality Controle
QMS	:	Quality Management System
QRA	:	Quantitative Risk Analysis
REPFED	:	Refrigerated Processed Foods of Extended Disability
RQ	:	Respiration Quotient
RTE	:	Ready to Eat
RTS	:	Ready to Serve
RTU	:	Ready to Use
SAM	:	S-adenosyl-Methionine
SFB	:	Solid Fiber Board
SMS	:	Sodium Meta-biosulphite
SO_2	:	Sulphur dioxide
SOP	:	Standard Operating Procedure
SQC	:	Statistical Quality Control
SQFC	:	Safe Quality Food Control
SR	:	Sulphur Resistant
SSOP	:	Standardized Sanitary Operation Procedure
SSP	:	Shelf Stable Products
ST	:	Storage Tastings
TBHQ	:	Tertiarybutylhydroquinone
TBZ	:	Thiabendazole
TDR	:	Thermal Date Rate
TDT	:	Thermal Date Time
TSP	:	Textured Soy Protein
TSS	:	Total Soluble Solid
TVP	:	Textured Vegetable Protein

UHT	:	Ultra High Temperature
VHT	:	Vapour Heat Treatment
VPD	:	Vapour Pressure Deficit
WHO	:	World Health Organization (UN)
WTO	:	World Trade Organization (1948), Geneva, (UN)
WVTR	:	Water Vapour Transmission Rate

❑❑❑

Contents

1

Importance, Scope and History of Post Harvest Technology

India is a second largest producer of fruits and vegetables after China. These share are 10% and 16% respectively in total world production. Horticulture crops occupy 23.548 million ha and production is estimated around 260.063 million tones. Total area of fruits, vegetable, plantation crops, spices, loose flower and aromatic plant are 7012, 9083, 3606, 3092, 218 and 537 thousand ha, respectively and production are 79400, 156445, 16388, 5786, 1421 and 622 thousand tones, respectively with productivity are11.32, 17.22, 4.54, 1.87, 6.59, 1.19 thousand tones per ha, respectively (2012-2013). The total productions of cut flowers are 74305 Lakh stem (2012-2013). The annual horticulture export volume is 2733145 metric tonnes (2010-2011) with total exported value of Rs 696461 lakh. In terms of volume, fresh fruits & vegetables comprised about 76 % of exported horticulture produce followed by processed fruits & vegetables (23 %) and floriculture & seeds (1%). After that losses found 20-30 % in fruit and 30-40 % in vegetable while harvesting to consumption in India. It's due to the perishable nature of horticulture commodity, non availability of appropriate post harvest infrastructure transportation, inadequacy of the market or lack of processing.

India is a major horticulture producer, even though processing is less than 2 %. This is very low as compare to 80 % of Malaysia, 78 % of Philippine, 70 % of Brazil and 30 % of Thailand. The most of developed country 40-60 % fruits and vegetables utilize in processing.

Proper handling, packaging, transportation and storage reduce the post harvest losses of fruits and vegetables. For every one percent reduction in loss will save 5 million tons of fruit and vegetable per year. Processing and preservation technology helps to save excess fruit and vegetable during the glut season. The technology has become a necessity to improve the food safety and strengthen nations food security. The technology helps to boost export of agricultural commodities in the form of preserved and value added products.

- Food processing industries could be divided into four group, based on their capacity of production:

  A. Large scale unit ›250 t/year

  B. Small scale unit 50-250 t/year

  C. Cottage scale unit 10-50 t/year

  D. Home scale unit ‹10 t/year

- Estimated post-harvest losses of fruit and vegetables

S. No.	Fruit/Vegetable	Percent loss
1.	Papaya	40-100
2.	Mandarin	20-95
3.	Banana	20-80
4.	Grape	27
5.	Lemon	20-85
6.	Cauliflower	49
7.	Tomato	5-50
8.	Onion	16-55
9.	Cabbage	37
10.	Potato	5-40

- Indian production of processed fruits & vegetables

Product	% Share
Fruit pulp & juice	27
Ready-to-serve beverages	13
Pickles	12
Jams & jellies	10
Synthetic syrups	8
Squashes	4
Tomato puree & ketchup	4
Canned vegetables	4
Others	18

- Indian processing industry profile

Industry	Share %
Unorganized	42%
Organized	25%
Small scale	33%

- India export of process products to different country

Product	Country
Mango Pulp	Saudi Arabia, UAE, UK, USA
Fruit Juice	USSR (Formerly)
Canned Fruit (mango, guava)	USSR (Formerly), UAE
Canned vegetable (Green pepper)	UK, USA, UAE
Dehydrated (Garlic, Onion)	USSR (Formerly), Japan, UK
Pickles, Chutney (mango)	USA, Japan, USA

- Factors affecting of post-harvest losses

  A. Before harvesting (Crops production) factor
    - Cultural operation (Training, Pruning, Staking or Weeding).
    - Mineral nutrients management.
    - Uses of PGR.
    - Water management.
    - Crop hygiene.
    - Chemical treatment.
    - Temperature.
    - Light Intensity.
    - Day length.

  B. Harvesting
    - Maturity.
    - Harvesting method.

  C. After harvesting treatment
    - Pesticides.
    - Fungicide.
    - Heat.

- Curing.
- Sprout suppressants.

D. Storage and transport conditions

- Packaging.
- Type of transport.
- Storage condition (Temperature, Humidity or Atmospheric gases).

History

- The first recorded cause of spoilage in stored food by Needham (1749).
- Iron container was discovered by Aes the bptngusts de Heine (1800).
- Peter Durand(1800) discovered the metal container.
- Fermentation firstly recorded by Lovoisier in 1789.
- Spallanzani (1765) disputed the theory of preservation by mean of heating. This constitutes the basic principle of canning.
- M. Nicholos Appart (1804) was the first to report the successful preservation of food in glass container. He is known as 'Father of Canning'.
- Fastier (1824) discovered the Hold the Cap Can.
- Cooking of food by mean of preserve started by Papin in 1861.
- Autoclave firstly used by Shriver in 1874.
- Modern refrigeration discovered by James Haryson in 1857.
- Fruit and Vegetable processing was Ist started in organized manner in 1857.
- In India, the first fruit and vegetable processing factory was established in 1935 at Mumbai.
- The Central Food Technological Research Institute was established at Mysore in 1950.
- Fruit Preservation and Canning Institute was established at Lucknow in 1949.
- Burg and Burg (1966) discovered the hypobaric storage.

2

Nutritive Value of Fruits and Vegetables

Fruits and vegetables are important part of our dietary, which are rich source of nutrient (carbohydrates, protein, fat, vitamins, minerals, and dietary fiber). Carbohydrates, protein and fat are macronutrients, considered as energy sources. Vitamins and minerals are micro nutrient, play important role in body building. Fibers have several direct and indirect advantages. ICMR, New Delhi recommends 125g of leafy vegetables, 100g of root and tuber vegetables and 75g of other vegetables (total 300g/day) and 85g of fruits per capita every day for balance diet.

1. Carbohydrates

- Carbohydrate is important and chief source of energy in human diet.
- Carbohydrates are classified in 3 groups.

Carbohydrates

Monosaccharides (Reducing Sugar)	Disaccharides (Non-Reducing Sugar)	Polysaccharides
(Glucose, Fructose and Mannose)	(Sucrose, Lactose and Maltose)	(Starch, Cellulose and insulin)

- 1g glucose liberates 4.0 calories of energy.
- Daily requirement of carbohydrate is 400-500g per capita.

Fruit Sources		Vegetable sources	
Raisins	77.3%	Cassava	38.1%
Apricot (dry)	72.8%	Sweet Potato	28.2%
Date	67.37%	Potato	22.6%
Karonda (dry)	67.1%		
Banana	36.4%		
Bael	30.6$		

2. Proteins

- Proteins are extremely complex nitrogen containing organic compounds. They constitute major part of protoplasm.
- Proteins, not considered to be primary body fuel, are also utilized for the production of organisms.
- Proteins made among 20 amino acids, in which 10 are essential amino acid as they are not synthesized in human body.
- 1 g protein liberates 4.0 calories of energy.
- Daily requirement of protein is 60-70g.

Fruit Sources		Vegetable sources	
Cashew nut	21.20%	Lima Bean	7.9 g/100g
Almond	20.88 %	Pea	7.2 g/100g
Walnut	15.60%	Cow Pea	4.3 g/100g

3. Fat

- Fat is stored energy sources of our body.
- 1 g fat liberates 9.0 calories of energy.

Fruit Sources		Vegetable sources	
Pecan nut	70.0%	Bengal Gram	1.40 g/100g
Walnut	64.5%	Potato	1.18 g/100g
Almond	58.9%	Small Bitter gourd	1.0 g/100g
Cashew nut	46.9%		
Avocado	22.8%		

4. Vitamins

- Vitamins classified in two group:-
  1. Water soluble: Vitamin B complex and Vitamin C.
  2. Fat soluble : Vitamin A, D, E and K.

Vitamin-A (Retnol, Carotene)

- Daily requirement is 1.2 mg/day.
- Deficiency symptoms: - Night blindness (Nyetalopia), Xeropthalmia for children, Keratnisation of epithelia cell of eyes.
- Vegetable are rich in 'Vitamin-A' than fruit.
- Carrot, muskmelon, winter squash and leafy vegetables are good source of vitamin-A.
- Carrot provide maximum vitamin-A per unit area.
- Precursor of vitamin-A is carotenoids.

Fruit Sources		Vegetable sources	
Mango	4800 IU	Bathua leaves	113000 IU
Papaya	2020 IU	Colocasia leaves	10278 IU
Persimmon	1710 IU	Turnip green	15000 IU
Date palm	600 IU	Beet leaves	9770 IU

Vitamin-B_1 (Thiamine)

- Daily requirement is 1.2 mg/day.
- Deficiency symptoms: - Beriberi, Muscular weakness, less of weight, Neuritis, loss of appetite and dilation of heart.

Fruit Sources		Vegetable sources	
Cashew nut	630 mg/100g	Palak	0.26 mg/100g
Walnut	450 mg/100g	Pea	0.25 mg/100g
Almond	240 mg/100g		

Vitamin-B_2 (Riboflavin)

- Daily requirement is 1.7 mg/day.
- Deficiency symptoms: - Dry scaly skin, crack in corners of mouth, cracking of lips etc.

Fruit Sources		Vegetable sources	
Bael	1191 mg/100g	Palak	0.56 mg/100g
Papaya	250 mg/100g	Chillies	0.39 mg/100g
Cashew nut	190 mg/100g	Fenugreek leaves	0.31 mg/100g
Pineapple	120 mg/100g		

Vitamin-C (Ascorbic acid)

- Daily requirement is 70 mg/day.
- Deficiency symptoms: - Scurvy.
- Approximately 90% vitamin-C is obtained from fruit and vegetable.

Fruit Sources		Vegetable sources	
Barbados cherry	1400 mg/100g	Drumstick leaves	250 mg/100g
Aonla	600 mg/100g	Coriander leaves	135 mg/100g
Guava	299 mg/100g	Chillies	111 mg/100g
		Broccoli	109 mg/100g

Vitamin-B_3 (Nicotinic Acid)

- Daily requirement is 19 mg/day.
- Deficiency symptoms: - Pellagra, nervous breakdown, stomach and intestinal disorder.

Fruit Sources		Vegetable sources	
Litchi	122.5 mg/100g	Palak	3.3 mg/100g
		Amaranths leaves	1.0 mg/100g

Vitamin-B_6 (Pyridoxine)

- Daily requirement is 1.3 mg/day.
- Deficiency symptoms: - Lack of Energy, Decrease in Brain Function, High Levels Of Homocysteine.

- Excellent *sources* of *vitamin B_6* include summer squash, bell peppers, turnip greens, shiitake mushrooms, and spinach.

Vitamin-B_{12} (Cyanocobalamin)

- Daily requirement is 2.4 µg/ day.
- Deficiency symptoms: - Pernicious anemia (Reduction in RBCs), Autoimmune disorders.
- Vitamin-B_{12} is providing by animal food. It is not found in vegetables and fruits.

Vitamin-D (Nicotinic Acid)

- Deficiency symptoms: - Rickets, Pigeon chest in children, Ostcomalacia (adult).
- Vitamin-D is synthesized by the body through sunlight.

Vitamin-E (Tocopherol)

- Daily requirement is 5.0 mg/ day.
- Anti-sterility vitamin.
- Deficiency symptoms: - Degeneration of kidney, Necrosis of liver.
- Sweet corn is rich source of vitamin E.

Vitamin-K (Phylloquinone)

- Anti-morrhagic vitamin.
- Daily requirement is 0.015 mg/ day.
- Deficiency symptoms: - Delayed and faulty coagulation of blood.

5. Minerals

Calcium

- Daily requirement is 500-600 mg/ day.

Fruit Sources		Vegetable sources	
Litchi	0.21%	Agathi	1130 mg/100g
Karonda	0.16%	Cury leaf	813 mg/100g

Iodine

- Daily requirement is 500-600 mg/ day.
- Deficiency of Iodine leads to goiter.
- Onion, garlic, beet, agathi leaves are good source of iodine.

Iron

- Daily requirement is 20 mg/ day.

Fruit Sources		Vegetable sources	
Dry karonda	39.1%	Amaranths leaves	22.9%
Date (Pind)	10.6%		

Phosphorus

- Daily requirement is 20 mg/ day.

Fruit Sources		Vegetable sources	
Almond	0.49%	Amaranths leaves	800 mg/100g
Cashew nut	0.45%	Garlic	187 mg/100g
Walnut	0.38%		

Potassium

- Spinach (605 mg/100g) and amaranths leaves (230 mg/100g) are good source of potassium.

Sodium

- Daily requirement is 4000-6000 mg/ day.
- Celeug, green onion, Chinese cabbage etc. are good source of sodium.

Folic Acid

- Daily requirement is 0.1mg/ day.
- Deficiency symptoms: - Anamia, nervous breakdown and impaired growth.
- Importance for women.
- Green leafy vegetables are good source of folic acid.

6. Fibers

Fiber is comprised of components of plant materials (long chain glucose molecules) that are resistant to human digestive enzymes. Fiber plays an important role in digestion, providing bulk, decreasing bowel transit time, absorbing toxins and cleaning the colon of mucus and undigested food particles. Fiber acts to increase the thickness of the stomach contents which gives a feeling of fullness and slows down the emptying of the stomach.

- Fibers are in two forms – soluble and insoluble.
- Vegetables and fruits are an excellent source of fiber, especially potato (2.2g/100g), broccoli, cabbage, cauliflower, corn, green beans, tomatoes, zucchini, fig, guava, apples, bananas, apricots, grapes, peaches, strawberries, etc.
- Maximum dietary fiber found in fig 7g/100g followed by guava (6.9%) in fruit and in vegetable potato have maximum dietary fiber 2.2g/100g.

Points to Remember

- One adult person requires 2800 kcal energy per day.
- In fruits, walnut (687 kcal /100g) have maximum calorific value followed by almond (655 kcal /100g), cashew nut (596 kcal / 100g) and date 283 kcal /100g and in vegetable, tapioca have maximum calorific value followed by garlic (142 kcal /100g) and lima bean (105 kcal /100g).
- Cobalt is present in 'Vitamin B_{12}'.
- Papin is prepared from papaya.
- Potato has 22% carbohydrate content.
- Grapes are rich source of tartaric acid.
- Vitamin-E is the anti-sterility vitamin.
- Carotene has xanthophyll hydroxyl group.
- Selenium (Se) is the anti-oxidant mineral.
- Avocado has 26.4% fat content.
- The fruits and vegetables are known as 'Protective Foods'.

- Protein content in potato is 1.4%.
- Ascorbic acid known as anti-oxidant vitamin.
- Vitamin-C is sensitive to heat.
- Fiber content in green chillis is 6.8%.

3

Maturity Indices and Harvesting of Fruits and Vegetables

Maturity

It is derived from Latin word *'Maturus'* which means ripen. It is that stage of fruit development, which ensures attainment of maximum edible quality at the completion of ripening process.

Maturation

It is the developmental process by which the fruit attains maturity. It is the transient phase of development from near completion of physical growth to attainment of physiological maturity. There are different stages of maturation e.g. immature, mature, optimally mature, over mature.

Maturity Indices

The factors for determining the harvesting of horticulture commodities according to consumer purpose and type of commodity.

Maturity Indices is two type

1. Physiological Maturity
2. Horticultural Maturity

1. Physiological maturity

Attainment of final stage of biological function by a plant part or the plant as a whole. (When a plant or plant parts will continue ontogeny even if detached).

2. Horticultural maturity

The stage of development of a plant or plant part, which possesses the necessary pre requisites for utilization by consumer for a particular purposes and it depends on the particular crop. e.g., apple, grape, mango etc. are harvested at fully developed stage but okra, bean, pot herbs etc. are harvest at tender stage.

Importance of Maturity Indices

- Ensure sensory quality (flavour, colour, aroma, texture) and nutritional quality.
- Ensure an adequate postharvest shelf life.
- Facilitate scheduling of harvest and packing operations.
- Facilitate marketing over the phone or through internet.

Ripe

It is derived from Saxon word *'Ripi'*, which means gather or reap. This is the condition of maximum edible quality attained by the fruit following harvest. Only fruit which becomes mature before harvest can become ripe.

Ripening

Ripening involves a series of changes occurring during early stages of senescence of fruits in which structure and composition of unripe fruit is so altered, that it becomes acceptable to eat. Ripening is a complex physiological process resulting in softening, colouring, sweetening and increases in aroma compounds so that ripening fruits are ready to eat or process. The associated physiological or biochemical changes are increased rate of respiration and ethylene production, loss of chlorophyll and continued expansion of cells and conversion of complex metabolities into simple molecules.

Senescence

Senescence can be defined as the final phase in the ontogeny of the plant organ during which a series of essentially irreversible events occur which ultimately leads to cellular breakdown and death.

Determination of Harvest Maturity Indices

1. Computational Methods
   a) Calendar Date
   b) Days from full bloom to harvest(DFFB)
   c) Mean heat units
   d) T-stage
2. Physical Methods
   a) Fruit retention strength
   b) Fruit size and surface morphology
   c) Weight
   d) Specific gravity
   e) Colour (Skin, flesh and seed)
   f) Flesh firmness
   g) Juice content
3. Chemical Methods
   a) Titrable acidity
   b) TSS/Acid ratio
   c) Total soluble solids (TSS)
   d) Sugars (Total and reducing)
   e) Sugar/Acid ratio
   f) Bioelectrical conductance
   g) Starch content
   h) Starch-Iodine test
   i) Tannin content
   j) Oil content

4. Physiological Methods
   a) Respiration rate
   b) Ethylene evaluation rate

Maturity indices for important fruits and vegetables

S.No.	Maturity Indices	Fruits/Vegetables
1	Calendar Date	All fruits
2	DFFB	All fruits and Radish
3	Mean Heat Units	Apple, pear, grape, mango, ber, litchi and many vegetables
4	T-Stage	Apple
5	Size	All fruits, beans, carrot, cucumber, cherry, asparagus and cauliflower
6	Surface Morphology	Grape (Cuticle formation), banana and litchi
7	Specific Gravity	Cherries, mango and ber
8	Fruit Retention Strength	Apple
9	Colour (Surface)	All fruit, muskmelon and tomato
	(Seed)	Apple, Aonla and pear
	(Flesh)	Mango, papaya, watermelon and muskmelon
10	TSS	All fruit, tomato and melons
11	Firmness	Pome and stone fruits, beans, lettuce and muskmelon
12	Juice Content	Citrus
13	Acidity and Sugar /Acid Ratio	Pomegranate, citrus, Grape, papaya and kiwi fruit
14	TSS/Acid Ratio	Grape and citrus
15	Sugars	Pome, stone fruits and grape
16	Astringency (Tannin)	Persimmon and dates
17	Oil Content	Avocado
18	Respiration and Ethylene Evolution Rate	Apple and pears
19	Aroma	Jackfruit

Maturity indices of fruits on the basis of TSS

S.No.	Fruits	Minmum TSS%
1	Appricot	10
2	Blueberry	10
3	Cherry	14-16
4	Grape	14-17.5
5	Kiwifruit	6.5
6	Mango	18-20
7	Muskmelon	10
8	Nectarine	10
9	Papaya	11.5
10	Peach	10
11	Pear	13
12	Pinapple	12-14
13	Plum	12
14	Pomegranate	17
15	Strawberry	7
16	Watermelon	10

Maturity indices of important vegetables

S.No.	Vegetables	Maturity Indices
1	Asparagus	Spear length (5-8 inches)
2	Bean	Uniformity in size and maturity and firmness
3	Broccoli	Compact flower heads, no open flower
4	Cabbage	Head solidity
5	Cauliflower	Head size and condition
6	Cantaloupes	Soluble solids (9%), ground colour, Netting
7	Carrot	Fruit size
8	Reddish	Days from planting
9	Celery	Size (Over maturity results and pithiness)
10	Lettuce	Firmness
11	Garlic and onion	Top dry, neck tissues begin to soften
12	Cucumber	Shape, size, colour and turgidity
13	Muskmelon	Soluble solids, firmness, shape, ground and flesh colour

Contd.

14	Watermelon	Weight(uniformity), flesh colour(75% red), soluble solids(10% or more), gives dull sound when tapped
15	Potato	Drying of foliage
16	Sweet potato	Senescence of vines
17	Colocasia	Leaves start drying
18	Tomato	Development of two or more seed cavities, ground colour
19	Brinjal	Immature, glossy skin, 40days from flowering
20	Pepper	Fruit size and colour (depends on colour and intended market)
21	Peas	Pods well filled but not faded in colour
22	Okra	Pod 2-4" long, not fibrous, tips of pods pliable
23.	Ginger	8-9 months after planting

Maturity indices for important fruits

S.No.	Fruits	Maturity Indices
1.	Almonds	Splitting of hull, separation of hull from shell, development of abscission zone
2.	Apple 'Golden Delicious' 'Red Delicious'	12% SSC, 18 lb firmness 11% SSC, 18 lb firmness
3.	Asian Pears	Skin colour change from green to yellowish green
4.	Pathar Nakh	145 days after fruit set
5.	Babugosha	135 days after fruit set
6.	Banana	Disappearance of angularity in a cross section of the finger
7.	Ber	Colour break stage (when light yellow colour appear)
8.	Cherry	TSS = 14-15%, light red colour
9.	Grapes Thompson SeedlessAnab-e-Shahi Bangalora Blue	TSS and TSS/Acid ratioTSS=18-22% and TSS/Acid ratio 25:1TSS=14-16% and TSS/Acid ratioTSS=12-14% and TSS/Acid ratio
10.	Guava	Colour break stage (when skin colour changes from dark green to light green)
11.	Lemon	30% or more juice by volume
12.	Lychee/litchi	TSS: acid ratio of 30-40, bright red in colour
13.	Kinnow	TSS/acid ratio 12:1 to 14:1
14.	Kiwi fruit	TSS – 6.5%, Firmness = 14 lbs
15.	Mango	Changes in shape (increase fullness of cheeks or bulge of shoulder), flesh colour yellow to yellowish-orange

Contd.

16.	Papaya	Skin shows yellowing
17.	Peaches	Ground colour change from green to yellow (varied for different cultivars)
18.	Plums	Skin colour changes
19.	Pomegranate	Minimum 1.85% TA and red juice colour
20.	Strawberries	2/3 of berry surface showing pink or red colour

Harvesting

Harvesting is a deliberate operation to separate the produce (economical part) from source of its attachment. This is very important operation as far as self life, transportation, marketing, processing, value addition and ultimately consumer satisfaction are concerned. Harvesting at immature stage and fallowing improper method lead damage to produce and producer both. Harvesting depend upon many factors such as nature and type of crop, delicacy of crop, maturity criteria, time of harvesting, mode of transportation, purpose of use, cost economy, etc.

Method of Harvesting

1. Manual Harvesting
2. Mechanical Harvesting
3. Chemical Harvesting

Some chemical are sprayed on the tree before harvesting to loosen the attachment of fruit to the plant. Following shaking the fruits fall down from the tree. Spray of 2,4-D @ 500 ppm and 200 ppm result in keeping the buttons of Kinnow mandarin and Eureka lemons fruit respectively, intact and green during storage.

Points to Remember

- Crane berry is first fruit crop which harvested by machine.
- Specific gravity of Alphonso mango has 1.01-1.02 and Dasheri mango has 1.0.
- Angularity or "fullness of fingers" appears is the standard maturity indices of Banana.
- Average sugar content in pineapple has 10-12%.
- Maturity indices for guava are specific gravity it has 1.0 and TSS has 12-13^0 Brix.

- Persimon have 14-17 % TSS level at the maturity stage.
- Juice content of ripe sweet orange has 35%, TSS -12^0 Brix and acidity 0.3%.
- Juice content of ripe mandarine has 33%, TSS- 12-14^0 Brix and acidity 0.4%.
- Juice content of mature lemon has 25%.
- Juice content of mature Grapefruit has 35%.
- Flattening of eyes is maturity Indices of Pineapple.
- Acidity percentage of mango is 0.5-0.6%.
- Acidity percentage of Pineapple is 0.5%-0.6%.
- Flattening of tubercles is the maturity indices of Litchi.
- Maturity of jack fruit judging by aroma.
- Specific gravity of potato is 0.98-1.02.
- Tomato are harvested at mature green stage for distant marketing.
- Tomato fruits for processing, are picked at hard ripe stage.
- Calliper grade is used for maturity measurement of banana.

4

Post Harvest Management of Fruits and Vegetables

Fruits and vegetables are living tissues subject to continuous change after harvest. Some changes are desirable from consumer point of view but most are not. Post-harvest changes in fresh fruit cannot be stopped, but these can be slowed down within certain limits to enhance the shelf life of fruits. The post-harvest handling plays an important role in extending the storage and marketable life of horticultural perishables.

Post Harvest Treatment of Fruits and Vegetables

Washing

It is useful in removing surface adhering material water. Mechanically, it is also accomplished by passing the fruits/vegetables through machine arranged with water tank equipped with rotating brush. In some produce washing improves appeal by facilitating removal of sap as case in mango, papaya, jackfruit, kronda, dirt in case of carrot, reddish, turnip, sugar beet, sweet potato, etc. and for removal of debris in case of banana. The water use for the purpose should be cleaned otherwise fungal and bacterial levels may build up. Disinfected treatment using ultraviolet light, ozone, chlorine (100 ppm), calcium hydroxide, etc. help to clean water for washing purpose.

Pre-cooling

It refers to the repid removal of the field heat from freshly harvested horticultural produce in order to slow ripening and senescence, conserve weight and reduce deterioration prior to storage and transport. Pre-cooling is accomplished by hydro cooling (immersing in water), contact icing (Placing cursed ice in or on the package), vaccum cooling (rapid evaporation of water at low presser to effect cooling) and air cooling (cooling based on refrigeration system).

List of commodity-wise cooling method

S.No	Cooling Method	Commodity
1	Room Cooling	All fruits and vegetables
2	Forced Air Cooling (Pressure Cooling)	Fruits and fruit type vegetables, tubers and cauliflower
3	Hydro Cooling	Stem and leafy vegetables, some fruits and fruit type vegetables
4	Package Icing	Roots, stem, some flower type vegetables, green onion and brussel sprouts
5	Vacuum Cooling	Some stem, leafy and flower type vegetables
6	Transit Cooling, Mechanical refrigeration	All fruits and vegetables
7	Top Icing and Channel Icing	Some roots, stem, leafy vegetable and cantaloupes

Curing

Curing done for root, tuber and bulb crops in which holding the produce at high temperature and high relative humidity for several days while harvesting wounds heal and a new, protective layer of cells form. While curing can be initially costly, the long extension of storage life makes the practices economically worthwhile.

S.No.	Commodity	Temperature (^0C)	Relative Humidity (%)	Curing Time (days)
1	Potato	15-20	90-95	5-10
2	Sweet Potato	27-33	>90	5-7
3	Yam	32-40	>90	1-4
4	Taro	30-35	>95	4-7
5	Cassava	30-35	>80	4-7
6	Onion and Garlic	35-45	60-75	0.5-1

Degreening

The process of decomposing the green pigment in fruits by applying ethylene (1000-2000 ppm) at 27°C and 80-90% or similar metabolic inducers to give a fruit to its characteristic colour as preferred by consumers, generally followed in citrus fruit but also practiced in banana, mango, tomato, etc.

Sorting

It is the process of separation of commodities into different categories on the basis of measurable physical properties. It also helps in removing of unwanted material or diseased portions.

Grading

Separation of commodity for getting maximum benefit on the basis of uniformity is termed as grading. It is an important aspect of quality as the product is rated on the basis of homogeneity of lot. Grading is done on the basis of size, shape, weight, colour, texture, firmness, composition, defects, etc.

Waxing

Waxing of fruits or vegetables is a common post-harvest practice. Food grade waxes are used to replace some of the natural waxes removed during harvesting and sorting operations and can help to reduce water loss and respiration during handling and marketing. It also helps in sealing tiny injuries and scratches on surface of fruits and vegetables. It improves cosmetic appearance and prolongs the storage life of fruits and vegetables. The wax coating must be allowed to dry thoroughly before packing.

Hot Water Treatment

Immersing the produce in hot water before storage/marketing is known as hot water treatment. It is effective in minimizing fungal diseases. Anthracnose in fruit can be controlled by dipping the fruit in hot water at 51-55^{0}C for 30 minutes. This treatment helps in attaining uniform ripening within 5-7 days.

Hot water treatments for different fruits

Commodity	Pathogens	Temp.(°C)	Time (min)
Apple	*Gloeosporium* sp. *Penicillium expansum*	45	10
Grapefruit	*Phytophthora citrophthora*	48	3
Lemon	*Penicillium digitatum Phytophthora* sp.	52	5-10
Mango	*Collectotrichum gloeosporioides*	52	5
Orange	*Diplodia* sp. *Phomopsis* sp. *Phytophthora* sp.	53	5
Papaya	*Fungi*	48	20
Peach	*Monolinia fructicola Rhizopus stolonifer*	52	2.5

Source: Lisa Kitinoja and James Gorny, 1998

Radiation Treatment

Subjecting the produce to ionizing radiation is referred to as radiation. It acts in pasteurization of surface. Hence, the technique is also known as radurization. As it works without raising temperature of produce, it is known as cold sterilization. It has its role in surface pasteurization, sprout inhibition and retardation of senescence process. Radiation can destroy pathogen/pest without raising the temperature of the produce.

Government of India approvals for Irradiation dose requirements for fruits and vegetables

Commodity	Irradiation dose (KGy)		
	Minimum	Maximum	Purpose
Onion	0.03	0.09	Sprout inhibition
Potatoes	0.05	0.15	Sprout inhibition
Raisins	0.25	0.75	Insect disinfestations
Dried fruits	0.25	0.75	Insect disinfestations
Mango	0.25	0.75	Shelf life extension and quarantine treatment, Fruit fly and stone weevil disinfestations
Ginger	0.03	0.15	Sprout inhibition
Garlic	0.03	0.15	Sprout inhibition
Small onions (shallots)	0.03	0.15	Sprout inhibition

Vapour Heat Treatment (VHT)

This treatment proved very effective in controlling infection of fruit flies in fruits after harvest. The boxes are stacked in a room, which are heated and humidified by injection of steam. The temperature and exposure time are adjusted to kill all stages of insects (egg, larva, pupa and adult), but fruit should not be damaged. A recommended treatment for citrus, mangoes, papaya and pineapple is 43^0C in saturated air for 8 hours and then holding the temperature for further 6 hours. VHT is mandatory for export of mangoes.

Fumigation

In this technique place the boxes of fruit in a gas (SO_2) tight room and introducing the gas from a cylinder to the appropriate concentration. However, special sodium metabisulphite pads are also available which can be packed into individual boxes of a fruit to give a slow release of SO_2. The primary function of treatment is to control the *Botrytis Cinerea*. The SO_2 fumigation is also used to prevent discolouration of skin of litchis.

Fumigation with 1.2% sulphur dioxide for 10 minutes was shown to be effective in reducing skin discolouration in fresh litchis, especially if it is combined with a 2 minute dip in HCl acid directly afterwards. Immediately after sulphur dioxide treatment litchi fruit may appear a uniform yellow colour and then turn red again after 1 or 2 days.

Chemical Treatment

- 100 ppm chlorinated water for fruits and vegtables disinfection.
- The fungicide, chemical and growth regulators are used to extend shelf life of fruits.
- Pre harvest treatment of Maleic Hydrazide (200 ppm) reduces sprouting of onion and potatoes during storage.
- Potassium permanganate is used as ethylene absorbent in packaging of fruits which delay ripening.
- Post harvest dip with Diphenylamine (0.1-0.25%) for 30 seconds controle superficial scald.
- Ripening of banana and tomato can be delayed by post harvest treatment with Gibberellic acid.
- Pre harvest spray of N-Benzyladenenine(BA) 10-20 ppm prolong shelf life of vegetables.

- Maleic hydrazide @ 1000-2000 ppm delays ripening of mangoes.
- Pre-harvest spray of calcium chloride (0.6%) and calcium nitrate (1%) resulted in enhancement of shelf life of mango and guava fruits.
- Treatment of mature green guava fruit with GA_3 at 100-200 ppm help extends the storability of fruits.
- Pre harvest treatment of ber with Calcium compounds (1.7%) solution is useful in delayed fruit ripening.
- Storage life of leafy vegetables can be extended following post harvest application of Cytokinin.
- Treatment of legumes vegetables with Indol Acetic Acid (IAA) solution is useful in maintaining green colour.
- Pre harvest spray of 0.2% Difolatan control post harvest diseases of tomato and onion.

Points to Remember

- Use of 4% wax emulsion useful is extending shelf life of mango fruits.
- Fruits of custard apple could be stored for a week after treatment with 8% wax emulsion.
- Wax emulsion at 6% enhances shelf life of litchi fruits.
- The rate of water loss can be reduced 30-50% under commercial conditions by waxing.
- Paraffin, carnauba, shellace and polythene based waxes are in use to improve appearance and self life of fruits.
- Artificial curing of onion is done at 40^0C temperature.
- Wax having two type (wax 'O' and 'w') both content 12% TSS.
- Wax 'O' impart gloss to fruits vegetables and wax 'W' dose not impart gloss to fruits vegetables.
- Hydrocooling at 12-15^0C with 500 ppm Bavistin increase shelf life of mango.
- The forced heated air is used for curing onion and other bulb, one day or less at 35-45^0C and 60-75% RH is recommended.

- In degreening CO_2 level below 1% does not allow higher colouring.
- Sweet potatoes are kept for 10 days at 80°C and 30% RH for curing.
- Best degreening temperature is 27^0C and Relative Humidity is 85-90%.
- Sprouting of potato, onion, garlic, tuberose, dahlia, crocus, etc. can be prevented by ionizing treatment at 10-15 K rads.
- An irradiation dose of a 10^2-10^3 rad is kill microorganisms and 10^3-10^6 rad insects.
- Ripening of banana can be delayed following irradiation at 25-35 K rads.
- Grape is first commodity which applied grading and marketing rules.
- Vacuum cooling is most suitable for leafy vegetables.
- In pre-cooling, water is mostely removed by conduction.

❑❑❑

5

Post Harvest Management of Flowers

Flowers are highly perishable unlike other horticultural or agricultural crops. Owing to poor keeping quality the post harvest losses in floriculture are significantly higher than any other sector. Although there has been significant increase in the area, production and productivity of flower crops in the last two decades, there is an urgent need to minimize the huge post harvest losses in terms of the value of the produce which are estimated to be 30-40 per cent of farm value. The post harvest behavior of flowers is an outcome of the physiological processes, occurring in leave, stem, flower bud, leafless peduncle or scape connecting bud to the stem. Some of these processes may act independently to affect the senescence and vase life of cut flowers but most of them are inter-related. The nature and extent of post harvest damage is typical for each crop or cultivar. The post harvest losses become important especially when dealing with the export of fresh flowers to distant and foreign market. Therefore, patient, soft and expert handling of flowers is of utmost importance after harvest.

The post harvest quality of flowers depends upon mainly three factors.

1. Pre harvest factors
2. Harvest factors
3. Post harvest factors

Pre-Harvest Factors

Genetic or inherent makeup

Post harvest lasting quality of flower species and cultivars vary considerably due to differences in their genetic make-up. Gladiolus varieties White Prosperity, Sancerre, Suchitra, Eurovision, Nova Lux, Rose Supreme and Trader Horn possess the better vase-life.

Growing conditions

Most cut flower crops require well-lighted conditions. On the contrary, too high light intensities cause scorching and dropping of leaves and abscission of petals. Flower crops are also specific in their temperature requirements. Flowers also require adequate nutrients for good longevity. High nitrogen doses should be avoided as they increase susceptibility to diseases. For example, iron deficiency is commonly observed in gladiolus in north India, causing heavy yield losses. Flowering crop should also be grown away from the industries which release toxic effluent, gases, damaging the foliage as well as flowers. Flowers damaged by pathogens, insects and pests also show high ethylene production resulting in poor vase-life.

Harvest factors

The most important factors for harvest are when, how and where, "when" the plant material will reach the optimum stage of development and "when" during the day to harvest. Each plant material has its own best harvest stage and this can vary depending on the use of, and market for, the plant material. Materials for preserving usually are harvested more mature than those for fresh, wholesale markets. Some general rules of thumb for when to harvest are, spike type flowers harvest when one-fourth to one-half of the individual florets are open; daisy type flowers-harvest when flowers are fully open. The other "when" is, the best time of day for harvesting flowers is coolest part of the day and when there is no surface water from dew or rain on the plants.

Right stage, method and time of harvesting of flowers are of considerable importance to ensure their long vase-life. The stems should be cut with sharp knives or secateurs. Hardwood stems should always be given slanting cut to expose maximum surface are to ensure rapid water absorption. The flowers of dahlia and poinsettia release latex upon cutting. To overcome such problem, stems should be given a dip in hot water (80-90^{0}C) for a few seconds.

The flowers of rose, carnation, gladiolus, tuberose, daffodils, lily, iris, freesia and tulip should be harvested at bud stage since their buds continue to open in water. The flowers of snapdragon, preffered to harvest at bud stage as their buds have long vase-life, less sensitive to ethylene, easy to handle during storage and transport as well as less prone to diseases and pests.

Harvesting stages for important Flowers

S.no.	Flower	Harvesting stages
1.	Rose	For local market: when outer one/two petals start unfurlingFor distant market: fully coloured tight budsLoose flower: fully open flower
2.	Carnation (standard)	For local market: when flower are half opened or at paint brush stage or outer petal is perpendicular to stemFor distant market: cross is developed on buds and colour is visible
	Carnation(spray)	For local market: when two flowers have opened and other heve shown colourFor distant market: when 50% flower have shown colour.Loose flower: fully open flower
3.	Chrysanthemum (standard)	For local market: half opened flowersFor distant market: when outer row of florets start unfurling
	Chrysanthemum (spray)	For local market: when two flowers have opened and other shown colourFor distant market: when 50% flowers have shown colour
	Chrysanthemum (loose flower)	Fully open flower
	Pot mums	50% buds have developed colour
4.	Gladiolus	For local market: lower most 1-2 florets are opened For distant market: lower most 1-2 florets show coloure
5.	Oriental lily	For local market: when 1-2 florets are opened For distant market: when 1-2 florets show coloure
6.	Asiatic lily	For local market: when 1-2 florets are opened For distant market: when 1-2 florets show coloure
7.	Gypsophilla	25-30% flowers are fully open in the inflorescence
8.	Tuberose	Single: buds are fully developed but yet not openDouble: basal 3-4 buds start to open
9.	Orchids	Most species: fully open flowersDendrobium: 75% inflorescence is open
10.	Anthurium	When one thred to one half of the spedix shows change in colour (mature)

Contd.

11.	Alstroemeria	For local market: when 4-5 florets have openedFor distant market: when first floret has started opening and other have developed 50% colour
12.	Gerbera	Before outer row of ray florets show pollen or when outer row of petals is perpendicular on stalk
13.	Marigold	Fully open flowers
14.	Calendula, pot marigold	Fully open flowers
15.	China aster, annual aster	Fully open flowers
16.	Cockscomb	One-half florets open
17.	Daffodil, narcissus, jonquil	"goose neck" stage
18.	Dahlia	Fully open flowers
19.	Delphinium	One-half florets open
20.	Freesia	First bud beginning to open
21.	Goldenrod	One-half florets open
22.	Larkspur, annual delphinium	
23.	Lily-of-the-valley	One-half florets open
24.	Nasturtium	Fully open flowers
25.	Pansy	Almost open flowers
26.	Snapdragon	One-third florets open
27.	Sweet pea	One-half florets open
28.	Sweet william	One-half florets open
29.	Zinnia	Fully open flowers

Post Harvest Factors

Water relations

The termination of life of the harvested flowers depends on water uptake and transport, water loss and the capacity of the flower tissue to retain its water. A water deficit and wilting develop, when the transpiration exceeds absorption of water. The rate of water uptake of cut flowers depends on transpiration pull, temperature and composition of solutes . Disruption of water columns in stem vessels by air embolism and resistance to water flow in stems, also develop water deficit. Acidification of water and addition of wetting agent and flower food in the holding solution markedly improve water uptake of cut flowers.

Respiration

The rate of respiration depends on quantity of carbohydrates available in the harvested flowers, temperature and the use of certain chemicals to regulate it. With higher temperature, there is faster rate of respiration and burning of the tissue. Consequently, the life of flowers is shortened.

Relative humidity

It has, bearing on the transpiration rate. Higher the humidity in the air, less is the transpiration rate and vice-versa.

Growth regulators

Postharvest life of flowers can be controlled by growth regulators. Water relation changes associated with flower senescence are also influenced by growth regulators. Cytokinins delay senescence of some cut flowers. Depending upon the concentrations, GA in some cases promotes longevity of flowers, while this is also used in bud opening solution. The IAA promotes ethylene production of isolated carnation petals. In contrast, the senescence and abscission of poinsettia flowers is delayed by auxin.

Preservative solutions

Preservatives in the form of tablets or powder are prepared from a mixture of chemicals-sugars, germicides, salts and growth regulators. Various types of conditioners are sugar and biocide, antiethylene compound, and hydrated compound. The flowers like gladiolus, carnation, chrysanthemum and freesia are benefited most by the pretreatment. Antiethylene compounds in preservative solutions reduce the action of ambient ethylene as well as autocatalytic production of ethylene by fresh cut flowers. Fresh cut flowers responding to silver thiosulphate are carnation, orchids, gypsophila, gladiolus, gerbera, snapdragon, alstromaeria, agapanthus, anemone and sweet pea. Greatest improvement in cut flower quality and longevity is obtained when DICA or DDMH were combined with sucrose.

Pre-cooling and storage

Precooling is essential for removing field heat from flowers. This is done either by forced air cooling or hydrocooling to bring down temperature from 20^0-30^0C to 1^0C in a relatively short period. Other methods are room cooling and vacuum cooling. Flowers can be stored for a longer period at low temperature. There are two methods of cold storage-wet and dry. Wet method is short-term storage, in which cut stems are dipped in water. Dry storage is more labour-intensive method and costly. The controlled atmosphere based on reduction of respiration rates, conservation of respirable substrates during, storage, and delay in ethylene-triggered changes cause senescence. It involves the use of increased level of CO_2 and decreased levels of O_2 in the atmosphere, low storage temperature and prevention of the build-up of endogenous ethylene.

Packing and transporting

Lower rate of transpiration, respiration and cell division during transportation, are essential for long storage life and keeping quality. Before packing, flowers should be dried. They should be treated with systemic insecticides and miticides. Packing must ensure protection of flowers against physical damage, water loss and external conditions detrimental to transported flowers. Boxes made of corrugated fibre boards are good. Flowers sensitive to geotropic bending must be transported in an upright position. The flowers should be transported at an optimal low temperature. The relative humidity of the air during precooling and shipment of cut flowers should be maintained at the level of 95-98%. Lack of light during prolonged transportation particularly at high temperature causes yellowing of leaves in many flowers. Shipment of flowers is usually done by road, air and sea. For short distance and time period shorter than 20 hr, cut flowers may be transported in insulated trucks without refrigeration after precooling and proper packing. Air shipment is quickest and usually the temperature is not controlled during the flight the flowers should be pulsed with STS prior to air shipment

Post Harvest Treatment of Cut Flowers for Improve Shelf-life

Pre-cooling

Precooling is a treatment given to flowers to remove the field heat immediately after harvest. It can be done with ice cold water, cold water or forced air. Pre-cooling reduces respiration rate and decreases breakdown of nutritional and other stored material in the steam, leaves and patals, delays bud opening and flower senescence. It also delays rapid water loss and decreases flower sensitivity to ethylene.

Pre-cooling temperature of some important flowers

S.no.	Flower	Pre-cooling Temperature
1.	Alstroemeria, Gladiolus	4^0C
2.	Anthurium	13^0C
3.	Dendrobium	$5\text{-}7^0C$
4.	Carnation	1^0C
6.	Rose	$1\text{-}3^0C$
7.	Chrysanthemum	$0.5\text{-}4^0C$
8.	Cymmbidium	$0.5\text{-}4^0C$

Conditioning/ hardening

It restores the turgor of flowers wilted after harvest, storage or transport. Conditioning is done with dematerialized water supplemented with germicides and acidified with citric acid to pH 4.0-5.0. Some wetting agents like tween 20 @ 0.01-0.10% can be used for this purpose. Some chemical like STS, 8-HQC, sucrose etc can also be used for conditioning.

Impregnation

Loading of flowers with high concentration of silver nitrate or nickel chloride of cobalt chloride for a short period of time is known as impregnation. It protects the blockage of the water vessel in stem by microbial growth and decay and synthesis of ethylene. It is generally practice in crops like Gerbera, Carnation, Chrysanthemum and Gladiolus cut flowers.

Pulsing

Treating the flowers with high concentration of sucrose and germicide for a short period of time, in order to improve the shelf life and to promote flower opening. Pulsing is beneficial especially for flowers destined for long storage period or long distance transportation. The addition of sucrose in the vase water alone, may encourage increased growth of micro-organisms in the vase medium. Hence, the antimicrobial agents eg 8-HQ, 8-HQS, 8-HQC, silver salts, citric acid etc should also be incorporated to enhance better solution uptake that would suffice maximum effects of the supplied sugar.

- Sucrose is the main ingredient of pulsing solutions providing additional sugar, and the proper concentration range from 2 to 20%, depending on the crop. The solution should always contain a biocide appropriate for the crop being treated.

- Ethylene-sensitive flowers are pulsed with silver thiosulphate (STS). Treatments can be for short periods at warm temperatures (e.g. 10 minutes at 20^0C) or for long periods at cool temperatures (e.g. 20 hours at 2°C).
- Alstroemeria and lilies can be pulsed in a solution containing gibberellic acid to prevent leaf yellowing, and this is often a useful pre-treatment.
- Short pulses (10 seconds) in solutions of silver nitrate useful for some crops. China asters and maidenhair fern respond well to solutions containing 1000 ppm silver nitrate. Other flowers are damaged by these high concentrations, but respond well to 100-200 ppm (e.g. gerberas). The function of the silver nitrate is not fully understood. In some cases it seems to function strictly as a germicide (e.g. chrysanthemums). In all cases, residual silver nitrate solution should be rinsed from the stems before packing.
- Generally for rose and carnation, 5-8% sucrose solution is sufficient while for multi-floret spike like gladiolus and tuberose, high concentration of 10-20% is needed.

Bud opening

Use of germicides, sucrose and hormonal solution to promote the opening of immature buds in crops like chrysanthemums, rose, carnation, gladiolus, snapdragon, etc.

Grading, Bunching and Packaging

After harvesting the flowers should be graded according to various grades as per specification for local and distant market. Then these should be pulsed and made into bunches of 5, 10, 20, 50, 100. Cut flower should be packed in corrugated cardboard boxed or sleeves. Packaging must ensure protection of flowers against physical damage and for this cotton or news paper can be used as cushion material.

Storage

After pre-cooling and pulsing the flowers can be stored at low temperature i.e. in cold store to regulate the flower market or to avoid the glut in the market. Controlled atmospheric (CA) modified atmospheric (MA) or hypobaric (LP) storage method can be used to enhance the post harvest life of flower.

Recommended temperature and RH conditions for flowers storage

Name of commodity	Temp (°C)	RH (%)	Approximate Shelf-Life
Alstroemeria	1	90-95	1 week
Anthurium	13	80	1 week
Asparagus	0-1	95-100	7-10 days
Aster	0-1	90-95	1 week
Asiatic lily, Oriental Lily	0-1		
Baby's Breath, Gypsophila	0-1	90	3 week
Bird-of-Paradise	8	85-90	14 days
Carnation	1	90-95	2-4 weeks
Chrysanthemum	0.5-2	85-90	12-15 days
Crossandra	15-20		
Daffodil	0-1	90	2 weeks
Delphinium, Larkspur	0-1	90-95	7 days
Freesia	0-1	90-95	2 weeks
Gladiolus	7	90-95	7-15 days
Gerbera	2-4	90-95	1week
Leatherleaf Fern	1-6		2 weeks
Iris	0	90-95	1 week
Marguerite Daisy, Boston Daisy	0-1	90-95	3-7 days
Marigold	8-12	90-95	1 - 2 weeks
Rose	0.5-2	90-98	7-10days
Snapdragon	0-1		10 days
Statice	0-1	90-95	6 weeks
Stock	0-1	90-95	3-5 days
Sunflower	0-1		
Sweet Pea	0-1		1 week
Sweet William	2-3		
Orchid	5-7		2 weeks
Torch Ginger	12.5-15		10-17 days
Tuberose	7-10	80 - 90 %	5-10-days
Tulip	1	85	7 week

Transport

Flower should be transported in corrugated cardboard boxes. The flowers which are sensitive to ethylene, ethylene scrubbers containing $KMnO_4$ should be added to those boxes. Some of the flowers are like gladiolus and snapdragon are sensitive to geotropic bending, so these should be transported in upright position. Some of the flower crops show yellowing during transportation due to lack of light, therefore there should be a provision of light inside the transporting vehicle.

Fresh Flower Preservatives

Fresh flower preservatives are chemicals added to water to make flowers last longer. They contain a germicide, a food source, a pH adjuster, water, and sometimes surfactants and hormones. Germicides are used to control bacteria, yeasts and molds. These microorganisms harm flowers by producing ethylene, blocking the xylem, producing toxins and increasing sensitivity to low temperatures. Bacterial counts of 10 to 100 million per 1 milliliter impairs uptake, while counts of 3 billion per 1 milliliter causes wilting. Some common germicides are listed in table:

Germicide Types	Common Name	Recommended Concentration
8-hydroxyquinoline sulphate	8-HQS	200–600 ppm
8-hydroxyquinoline citrate	8-HQC	200–600 ppm
Silver Nitrate	AgNO3	10–200 ppm
Silver Thiosulfate	STS	0.2–4 ppm
Thiobendazole	TBZ	5–300 ppm
Quarternary ammonium salts	QAS	5–300 ppm
Slow-release chlorine compounds	50–400 ppm of Cl	
Aluminum sulphate	$Al_2(SO_4)_3$	200–300 ppm

8-HQC is the most common one used in commercial floral preservatives. Sucrose is the most common food source used in floral preservatives. It provides energy to sustain flowers longer and to open flowers in the bud stage. 1 to 2 percent sucrose is the standard amount in preservatives. Never use sucrose without a germicide, as it is the primary food source for microorganisms, too. Acids or acid salts are added to adjust the pH of the water to 3.5 to 5.0. At this pH, less microbes can grow and water is taken up by the flowers more easily. Surfactants and wetting agents reduce water tension so water is taken up more easily, also. Tween 20 and Triton are examples of surfactants.

Commercial Fresh Cut Flower Preservative

- Floralife
- Rogard/Silgard
- Chrystal
- Prolong
- Oasis
- Vita Flora

Holding

After pulsing and storage flowers are held in a solution containing sucrose, germicide ethylene inhibitor and growth regulator, that type of solution is called preservative or vase solution. The flowers can be kept in holding solution either at wholesaler, retailer or consumer level. Basically, the concentration is much lower than pulse solution. The anti-microbial agents like 8-HQC (8-hydroxy quinoline citrate), 8-HQS (8-hydroxy quinoline synthesis), aluminium sulphate, anti-ethylene agents like STS (Silver thiosulphate), $AgNO_3$ and acidifuing agents like citric acid are widely used as vase solution. Even plant growth regulaters like gibberellic acid, benzyl adenine, are also reported to improve vase life and quality of gladiolus and chrysanthemum. Growth retardants like CCC (cycocel), SADH (Succinic acid dimethyl hydraride) and MH (malic hydrazide) have also been reported to improve vase life of some cut flowers. Ethylene inhibitor like Amino ethoxy vinyl glycine (AVG), Methoxy vinyl glycine (MVG) and Amino oxyacetic acid (AOA) are beneficial in ethylene sensitive flower. Recently, chemicals like 1-MCP (1-methyl cyclopropene) and 1-OCP (1-octacalcium phosphate) have been found to be highly effective especially for ethylene insensitive flowers.

❑❑❑

6

Post Harvest Management of Spices and Aromatic Plants

Since time immemorial, spices have played a vital role in world food trade, due to their varied properties and applications. We primarily depend on spices for flavour and fragrance as well as colour, preservative, inherent therapeutic, medicinal and appetizing properties. The spices that India can offer in abundant quantities are pepper, ginger, turmeric, chili, cardamom, celery, fenugreek, fennel, cumin, dill, coriander, cinnamon, *ajowan*, cloves, nutmeg, mace etc.

- Spices add flavour, colour to food and make the food palatable.
- It stimulates salivation and acid secretion of digestive enzymes like amylase.
- It has anti inflammatory, antibacterial and antioxidant properties. It reduces cholesterol levels useful in preventing heart diseases etc.

Spices and their Uses

All spice or Pimento

It is a small berry, size of the pea, dried to a dark brown colour. The flavouring component is a volatile oil present at 3-4% levels containing eugenol and other closely related phenols. It is used whole in pickling and cooking meats and fish. It is used ground in cakes, puddings and preserves.

Ani seed (Somfu/owa)

The chief compound of it is anethole. It is also known as sweet cumin and has the flavour of liquorice. It is chewed after meals and used as mouth freshener. It is used in cakes, breads, cookies and candles; vadacurry, kurma, non-vegetarian dishes, biriyani, thandai, kachori and in pickles.

Asafoetida

It is an oleo gum resin exuded from the rhizome or root of Ferula asafoetida. The odour component consists of a ferulic ester and sulphur containing volatile oil.

Bay leaves (Birinj Leaves)

They are the dried aromatic leaves of laurel tree. They contain 1-3% of highly aromatic volatile oil. Oil of bay leaves is used in the preparation of pickling spice and in the flavouring of vinegar.

Caraway seeds

It contains about 5% volatile oil. The chief flavouring principle of which is D-carvone and D-limonere. The seeds are used to flavour cakes, biscuits, cheese, apple sauce and cookies. The oil from caraway seed is used to flavour meats, canned foods, sausages, soups, in cakes, certain bread rolls, cheese, confectionery and also flavouring liquor.

Cardamom

The fruit contains brownish black seeds, which have about 2-10% volatile oil with the characteristic pleasant odour. The active principles present in the oil cineole, terpinyl acetate, pinene, sabinene and porneol.

Chillies

The red colouring matter of chillies is due to a carotenoid pigment (capasin). Ground chilli is used in most of the Indian gravies and vegetable dishes. It is used in making chutneys, pickles, dehydrated chillies, vattal and in day to day preparations. Dry chillies are used for seasoning. Chillies contain a substance called capsaicin, which increases the gastric secretion, and it causes the destruction of the mucosal cells.

Cinnamon

It is a thin inner bark of a cinnamon tree. The bark contains about 1% essential oil. The active principles are eugenol, cineole and cinnamaldehyde. It is used in stick form in fruit preserves. It is used ground for cakes, cookies and puddings. Used in spicing sauces,pickles,in pulav,biryani.

Clove

It is the dried flower bud of the clove tree. It contains about 15% essential oil. The chief active principle is eugenol. Cloves are used whole in meats, pickling and fish. It is used ground in cakes, cookies and puddings. Used as flavouring agent in pulav, rice puttu and fruit cake.

Coriander seed

It is unmatched for its fresh delicate spring like aroma. The seeds contain 0.5-1% essential oil which has an active principle coriander an isomer of geraniol. The roasted powdered seed are used as an ingredient of curry powders. It is used as flavouring and thickening agent in cookery. In the form of powder it is used in rasam, all curries and vegetables and chutney powders. It is also used as whole in kamandhokla, samosas and kachori. It is used for sprinkling on prepared food, flavouring sambar and rasam, pounded or minced as a base for sauces or chutneys.

Cumin seeds

It is pungent, sharp and astringent. It contains 2-4% essential oil. The active principle is an aldehyde cumino.

Fenugreek seeds

It is a hard lentil seed. Its colour is dark fawn and has astringent aroma Fenugreek seeds contain 5% bitter fixed oil. The oil has a strong celery like odour.

Garlic

Garlic contains an antibiotic principle "allin" (inactive form) which is converted to allicin (active form) by the enzyme allinase. Allicin further breaks down to allyl disulphide, which is responsible for characteristic flavour. It is used in reciepes like rasam, pickle, chutney, pulav and sauces.

Ginger

It is the root of the plant *Zingiber officinale*. The volatile oil present is "gingerol". The flavouring compound has sharp burning sensory stimulation.

Kokam

It is a fruit dried and used as sour agent in cookery. Since anthocyanin is present, it is also used in making sherbet.

Mango powder

During the early stages of growth, the tree may have a heavy "fruit drop" that is use for making of mango powder. It is rich source of vitamin-A.

Mint leaves

It is essentially an aromatic culinary herb. These are the leave of spear-mint plant. The leaves yield an essential oil (pepper mint oil) which is used for flavouring gum, confectionery, tooth paste, perfumes and pharmaceutical preparations. The principle components of the peppermint oil are menthol, menthyl acetate, menthyl isovalerate and menthone. It is used fresh for beverages, salad dressing and for garnishing. It is also used in making many dishes like raitha pulav, chutney, vada and pani puri water. Dry powder is also used for different purpose.

Mustard

These are the small reddish black seeds of annual herbs. The leaves of this plant are consumed as vegetable. Mustard seeds have a pungent flavour. The characteristic flavour of mustard seed is due to an allyl iso thiocyanate. The mustard paste is used to flavour hotdogs, sandwiches, cheese, eggs, meat and salad dressings. Dry mustard is used in meat, sauces, gravies and mayonnaise. It is used in the powdered form in pickles. It is used in the vegetable preparations and raitha. It is used mainly for the seasoning. Mustard is used without the skin for pickling (chilli pickles). Mustard oil is also used in cookery.

Nutmeg

Nutmeg is the dry hard wrinkled seed or pit of the nutmeg fruit. Mace is the orange red fleshy covering of the nutmeg kernel. They contain 7-14% essential oil. This essential oil contains highly toxic compound known as myristicin, used in small amounts to flavour puddings and fruit Myristicin present in nutmeg seeds could lead to delirium and deep stupor. Smaller amounts of nutmeg may cause vomiting and colic.

Mace

Mace is more delicate and used for flavouring fish sauces, certain meat and fish dishes, pickles and preserves. It is used ground in cakes, cookies, pies and chocolate dishes. It is used in garam masala.

Onion

It is used as a flavouring agent in food preparations. It contains an essential oil the active principle of which is responsible for the characteristic cooked flavour is allyl propyl disulphide. Dehydrated onion is used in European countries and U.S.A. for flavouring food preparations. It is used in cookery to improve the flavour or to mask undesirable flavours and to increase the thickness of the gravies.

Omum

It is allied to the liquorice family and a powerful spice. It is used in ompodi, rusks, namkin and biscuits.

Poppy seeds

They are the tiny dark cream colour seed of poppy plant, which are used for topping of bread, cake, rolls or fillings for buns. Oil extract is used for salads. Poppy seeds ae used as thickening agent in the preparation of gravies. It is also used in kurma, non-vegetarian dishes and sweets.

Pepper

They are the dried small round berry of a tropical vine with small white flower. White pepper is mature berry with the black coat removed. Pepper corns are used whole or crushed. Ground pepper is used for seasoning many dishes and is also used as a condiment at the table. Pepper corns are generally used whole in pickling meats and stews. It is used ground for general seasoning of meat fish, poultry, vegetables and salads. White pepper is used in dishes that require a less pungent flavour. It is used in making bounder, pongal, rasam, kolambu, vadai and fried rice. And also used as substitute for chilli powder. It is also used in marinating the non vegetarian foods. Flavour can be improved by the addition of pepper powder to omelets, sandwich, salads, papads, soups and chips.

Saffron

This is the name given to the fragrant stigmata found in the Crocus flower. It is a regal spice of matchless aroma and the most costly in the world. 75,000 flowers are needed to make one ounce of pure saffron. Each filament can colour 700 times its own weight in water. It is used mainly for its yellow colour. Saffron has a pleasant aroma and an essential oil croncin and the colouring principle crocerin. It is used in soups, sauces especially rice dishes to give them bright yellow colour and distinctive flavour. It is used in sweets, like sandesh, rasmalai, thandai, kesar milk, ice cream, halwa and srikand. It may be adulterated with styles, anthers and parts of corolla of saffron and various materials coloured with coal tar dyes.

Tamarind

This spice is the pulp of the tamarind fruit after the outer shell and seeds are removed. The extract is used in making rasam and sambar. It is used as souring agent in chutneys, chat, pickle, pani puri and in tamarind rice. It is also used as thickening agent in gravies. Tamarind puree is available in the market.

Turmeric

It is the ground dried aromatic root. It contains 5% essential oil. The colouring substance present is known as curcumin. Its natural aroma is most apetising and on its own without the use of other spices, can produce delicious food.

Vanilla

Vanilla beans are being cultivated in Kerala. The active principle is "vanilla". Synthetic vanilla is much cheaper than that of the natural product. It imparts the flavour and it is used in preparing ice creams, custards, puddings and cakes.

Harvesting

Plants should be harvested during the optimal season or time period to ensure the production of plant materials and finished spice products of the best possible quality. The time of harvest depends on the plant part to be used that are under:

Harvesting Seeds and Fruits

Maturity

The selection of seed or fruit to harvest which is at the correct maturity or ripeness, based on colour, is critical to obtaining a product of even high quality.

Allspice (Pimento) fruit is harvested 3-4 months after flowering when mature but not ripe (green) fruit is most strongly flavoured. Small or over-ripe berries should be removed as they detract from the appearance of the finished product. Harvesting is undertaken manually by breaking off the twigs bearing the berry clusters. The berry clusters should be placed in a container and not placed on the ground, as soil contamination aids microorganism growth. Leaf oil is also produced from the material collected during the berry harvest or from specialized plantations.

Vanilla beans are ready to harvest 6-9 months after flowering. Colour is an important indicator for vanilla pod maturity and daily picking is advised to ensure pods are at the right stage, as immature pods produce an inferior product and overmature pods split during curing. The right picking stage is when the distal end of the pod turns yellow and fine yellow streaks appear on the pods. Vanilla pods, when green, are almost odourless but they develop a faint phenolic odour as the beans reach a stage of harvest maturity.

Coriander harvest time is of paramount importance in determining the aroma quality of the spice. Immature fruit contain higher volatile oil than ripe fruit but have a disagreeable aroma. The distinctive sweet spicy aroma of coriander spice doesn't develop until the fruit reaches maturity and begins to dry. The normal harvest time is when the fruits on the primary umbel reach a chestnut colour. If fruits are allowed to over ripen there is a high risk of shattering during harvesting, cleaning and drying.

Harvesting Leaves and Stems

The aromatic herbs such as basil, bay leaves, mints, oregano, parsley, rosemary, sage,tarragon and thyme are often used fresh to garnish food, as dry seasonings or for extracted essential oils. The aroma, a function of their essential oil composition, is dependent on chemotype and specific chemotypes can be chosen for an end-use. The essential oil composition and yield will also change during the seasons, and crop harvest should be directed to maximise both the essential oil content and quality. As a general rule, the leaf and stem should be cut when the growth has matured to an elongated flower stem but without full flower or significant senescence of the lower leaves. The essential oil can be distilled from fresh or dry plant material. Harvesting the plant material should be avoided when it is wet and if it is to be dried the plant material should be evenly spread onto the drying racks or drier to ensure there is no sweating, fermentation and microbial invasion.

Harvesting Roots and Rhizomes

The common spices which are derived from roots, bulbs and rhizomes are turmeric, ginger, onion, garlic and horseradish.

Turmeric, a perennial herb grown as an annual crop, is ready for digging when the lower leaves turn yellow. The length of time to harvest maturity is dependent on cultivar. Hand digging is the most common form of harvesting the rhizomes although diggers and lifters can be used. The leaves must be cut prior to mechanical lifting or after hand digging. Care is necessary to avoid damage to the rhizome (splitting or bruising) as injuries can result in fungal infection and rejection. Rhizomes are lifted whole, washed, sun dried and the rhizome fingers (which are kept as seed material) are separated from the mother rhizome.

Ginger is a perennial herb but is often grown as an annual. Its harvesting depends on the cultivar and varies from 7-9 months for annual crops while perennial crops are harvested at the growers' discretion. The time to harvest is also dependent on the projected end-use with fresh ginger with low fibre content harvested at about 190 days after planting. The harvesting method must ensure there is no rhizome damage and both manual and mechanical methods of rhizome digging are used. Some production areas use both methods by mechanically loosening the rhizomes to assist the hand lifting. Following lifting the rhizomes are washed, the roots removed and then killed by immersion in boiling water for about 10 minutes, dried and then stored.

Onion harvest time and maturity is based on top senescence when the bulb is converted from active growth to dormancy by under cutting that begins the drying process.

Garlic can be harvested at different stages of maturity for specialty markets but most commonly when the bulbs are well mature and the tops have dried completely.

Horseradish, a perennial that is grown for its enlarged root, is ready for harvest after the leaves have been killed by frost. Horseradish that is harvested while it is still actively growing does not store as well.

Harvesting Bark, Wood and Resins

Many plants which are harvested for wood or bark products, are managed under coppice plantation systems.

Cinnamon and **Cassia** production is an example of such a practice. Harvesting of the shoots is undertaken during the rainy season two years after coppicing. The selected shoots must have a uniform brown colour of bark and have at least two years growth. The rough outer bark of the selected shoots is first scraped off and the young tender inner bark is peeled off carefully from the stem. The inner bark curls naturally into the well-known quills. The best pieces of the peeled bark are packed with small pieces and tightly rolled to preserve the flavour and then dried. The coppiced shoots are left for fermentation for 24 hours, dried in the shade for one day and in the sun for four days. The smaller quills are inserted into larger ones to form compound quills. The products are graded as quills, quillings, featherings, scraped chips and powders. The finest quality bark is obtained from shoots (<1.25 m by 1.25 cm diameter) with uniform brown thin bark harvested at six-month intervals.

Drying

This the most critical process in the production of dried herbs and spices. The aim of drying is to reduce the moisture content of the product from actively growing in the field to a level that prevents deterioration of the product and allows storage in a stable condition. Drying is a two stage process: firstly the transfer of heat to the moist product to vaporize the water in the product and secondly mass transfer of moisture from the interior to the product surface where it evaporates.

There are four main types of drying. The most basic method of drying is to spread the crop on a surface exposed to the sun. In this case, the process is aided by a cover system that prevents wetting with rainfall.

An improved method, speeding up the drying, is to use a fuel source (wood, oil/diesel, gas or electricity) to heat the drying room. Solar drying systems together with solar powered fans are also available. The drying process should dry the crop as quickly as possible, at temperature levels which do not drive off the volatile flavor compounds. The drying temperature regime will be specific to each crop as will be the final moisture percentage for storage.

Drying seeds and fruits

Pepper, after threshing and blanching, the fruits are dried in the sun for 7-10 days till the moisture is about 10% and the seeds have the characteristic wrinkled appearance of black pepper. Pepper drying in the sun has to be turned over periodically to aid uniform drying and discourage mould infestation. The main disadvantage of sun drying is the lack of uniformity and potential contamination by microorganisms. Improvements in black pepper need to focus on the drying process. In areas where there is rainfall all year round and a high humidity, the quality of black pepper produced is dependent on drying. Mechanical driers have only had limited success at farm level and the cost of drying using fossil fuels is relatively high but other forms of drying such as solar-driers or solid fuel driers have not been found as effective.

Nutmeg & Mace fruit are dried in their shells in the sun and turned each day to prevent fermentation. The nuts are sufficiently dry when they rattle and drying takes about one week. Artificial dryers are sometimes used. Once dried the nuts can be stored for a considerable time. The separated mace is flattened by hand and dried on mats in the sun. This takes between 2-4 hours. In Grenada mace is cured in dark storage for 4 months. This produces a brittle pale yellow mace that has a premium price.

Chillis (Capsicums, Chilli peppers and Paprika) have a thick waxy skin, which prevents rapid drying. Sun drying of fruit is the most common method and is undertaken by spreading the crop in a dry area exposed to the sun or under cover. Weather exposure can bring high losses and contamination is also a problem. Fruit are spread in thin layers on a hard dry surface. The use of black plastic under the fruit speeds the drying time. The fruit is regularly stirred to ensure uniform drying and to reduce discolouration and fungal growth. At night or in wet conditions the fruits are heaped and covered. Drying will take up to 15 days dependent on the sunshine hours and weather conditions. A 100kg of fruit produces about 25 to 35kg of dried spice. Large traders use racks that can be moved in or out of the sun until the fruits are dried below 10% whilst retaining colour and pungency. This system reduces the drying time by half and produces a more uniform product. A chemical emulsion 'Dipsol' can be sprayed on fresh fruit to accelerate sun drying. Artificial driers such as energy-efficient heat pump dryers are well suited for drying chillies as they operate at low temperatures. This can be essential for chilli drying under changeable weather conditions as the spice quality becomes less desirable when it becomes brown instead of bright red. The drying temperatures should be below 60°C.

Vanilla pods, during the curing process, develop flavour as a result of naturallyinduced enzymatic action. The final vanilla quality depends mostly on the glucovanillin content in the green bean. This means that the quality of cured beans will be higher if the curing starts with mature beans that have the highest glucovanillin concentration. A number of procedures have been developed for curing vanilla, and they can be characterised into four stages. The first operation is to halt the plant's physiological functions by stopping vegetative development so that various enzymes can come into contact with cell substrates. The most practical methods are sun- or oven-wilting or hot water scalding. The second operation that is the most crucial step is sweating. Bean moisture is initially allowed to escape rapidly to get to a level that will reduce the risk of microbial spoilage during the subsequent processing. The curing enzymes are most active during this stage and vanillin and related compounds are released during this step that takes 7-10 days. At the end of the sweating process, the cured beans are still 60-70% moisture that is too high for safe storage. The third operation is drying to protect the beans from microbial spoilage and also to inhibit further enzyme activity and biochemical changes. At the end of drying the beans should be at 25-30% moisture. The fourth step is storage in closed boxes for several months while chemical reactions take place. There are two main methods of curing, the Mexican process (sun method) or the Madagascar process (Bourbon method). The Mexican method begins with shed storage for some days until the beans start to shrink. The beans are then placed in sweating boxes, for at least 24 hours, where the temperature is maintained to encourage the slight fermentation and flavor developing enzymatic reactions. This process of wilting and sweating may be repeated a number of times during the early part of curing and extend over a 2-3 week period. The moisture is reduced to the desired level of 25% by frequent exposure to sunlight and sweating. The final curing occurs when the beans are matured in boxes for 2-3 months. The Mexican process of curing takes 5-6 months. The Madagascar process is faster and begins when the pods are dipped in hot water at 80°C for 7-15 min. The beans are then spread out on blankets in the sun, and when well heated are rolled in blankets to stand and ferment overnight. This procedure is repeated over 10 days. At the end of this period the beans are placed on trays and allowed to further cure and dehydrate. The Madagascar process takes 35-40 days.

Allspice - Pimento berry clusters are taken to a drying shed and left in heaps or sacks for up to 5 days to ferment. Berries are then spread on outdoor drying floors, and turned daily to ensure even drying. The length of drying is dependent on the weather conditions (5-10 days) and the berries are covered at night and during rain. Moisture content close to 12% is the target and the yield from 100kg of green berries is about 55-65kg. Microbial contamination often occurs during the fermenting and drying phase, especially during wet and humid weather. This contamination is almost impossible to counteract and a high microbial count results in berries unsuitable for human consumption. These berries can be distilled. Correctly dried spice should have a pleasant characteristic odour and low microbal counts. These objectives are most easily met by artificial drying, under controlled conditions. Artificial driers (wood fired) are more common in areas where berries ripen in the wet season. Solar driers are more efficient but require a higher standard of post-harvest berry management. Permanently-placed large forced air driers owned and operated by grower cooperatives or local government, are becoming more common. A maximum drying temperature up to 75°C can be used without loss of essential oil. The problems of microbial contamination can be kept to low levels with good drying and storage management. The dry berries are cleaned and bagged, and kept in a clean dry store.

Cardamom capsule colour and the retention of greenness are important for highest quality. Capsules are cured immediately after picking to retained greener colour. The loss of greenness was more significant if the capsules were stored for more than 12 hours. Bagging of the capsules in jute bags, and cool storage, aids the retention of greenness. Cardamom capsules at picking contain 70-80% moisture and to enable storage the moisture has to be taken down to 8-10% by curing. Sun-drying is generally undesirable for cardamom because of bleaching and capsule splitting. The most widely-adopted drying system is a slow dry over 18-30 hours using a number of methods of artificial drying (electric, kiln, bin) with various methods of hot air flow. A temperature between 40- 50°C helps retain the green colour and an increase in temperature significantly increases the percentage of yellow capsules, split capsules, and heat injury.

Drying leaves and stems

Drying is the transformation of the harvested leaf and stem, containing 80-90% water, to a stable state containing 5-10% water. It is important to minimise the holding time between harvest and drying, and these activities should be co-ordinated to prevent delays in drying and the chance of spoilage. The temperature during drying is critical and if the drying temperatures are too hot the volatile components in the essential oils will be depleted and lost. In general drying temperatures should be below 40°C with forced air movement.

Drying flowers and buds

Clove buds are spread onto matting and dried in the sun. Occasional gentle raking will aid even drying and it is important to overcome heating and mould formation. Damaged and spoiled buds are removed during this phase. Drying is completed in 4-6 days with buds losing about two-thirds of their weight.

Saffron stigma should be dried the same day as picked from the field, following separation from the floral parts. The stigma is about 85% moisture at field picking and to enable long-term storage of the filaments, a moisture level close to 10% is required after drying. To obtain high quality saffron, drying temperatures should initially be high (<90 °C) for a brief period (< 20 minutes) and then continued drying at a lower temperature (d"40 °C). Irrespective of the drying method, it is important not to over-dry to a brittle stigma. The duration of drying is dependent on the drying temperature used and the higher the temperature used, the shorter is the drying time.

Drying roots and rhizomes

Turmeric rhizomes have to be cured after harvest for both colour and aroma. The traditional method of curing rhizomes is to steam or boil fresh rhizomes in lime or a 0.1% sodium bicarbonate solution. The 'curing' is to remove the raw odour, reduce the drying time, gelatinise the starch and produce a more uniformly-coloured final product. Traditionally the cooked rhizomes were spread in the sun to dry and this process takes 10- 15 days. The final moisture level should be close to 6% moisture content. Today the majority of internationally traded turmeric rhizomes are artificially dried with hot air drums, tray and continuous tunnel driers and in India a maximum temperature of 60°C is advised. An important factor in drying time is the preparation of the rhizome. Rhizomes to be dried can be sliced or whole, with slicing generally producing a more uniform and brightly coloured powder. The yield of the dried product varies from 10-30% depending on the variety and the crop-growing environment.

Ginger curing of the rhizome prior to drying directly affects the fibre and volatile oil content. Removal of the skin reduces the fibre content and also increases the oil loss. Peeling also affects the pungency as these compounds (gingerols) are in the skin. When sun drying is not an option wood-fired or solar driers can be used while in Australia gasheated dehydrators are used. The final dry matter should be in the 7-12% range with a weight loss during drying of 60-70%. Artificial drying minimises the loss of quality and can also eliminate microbial contamination. Drying temperature, airflow and the length of drying all affect the flavour compounds in ginger.

Garlic subsequent to the field harvest, the bulb is broken into individual cloves and the loose paper shell removed by screening and airflow. The cloves are then washed and sliced. The sliced garlic cloves are dried down to about 10% moisture, with a drying temperature below 60°C. In the final drying procedures care is needed to ensure there is no heat damage to the slices and the garlic is dried to 6.5% moisture.

Onions can be dehydrated in the form of flakes, rings, kibbles and powder. Small-scale onion drying has been achieved by solar drying, but cabinet drying at 55-60°C for 10-15 hours gives a better product than sun drying or drying in solar huts. Commercial dehydration is achieved by forced hot air with an initial temperature of 75°C reducing to 55- 60°C as the moisture content falls. Final storage moisture content close to 4% is sought.

Horseradish are very sensitive to wilting and immediately after harvest the roots should be pre-cooled to <5°C using forced air cooling or placed at 0°C with 90-98% RH. Roots can be stored for 12 months under those conditions but pungency is rapidly lost at higher temperatures and roots dry out at lower RH. Perforated polyethylene bags and lined crates can maintain high RH during storage.

Cleaning

Cleaning the spice prior to packaging and sale is to ensure that the spice is of the highest quality and will obtain the highest price. Cleaning should remove all the foreign matter that lowers the quality and endangers the sale. Sieves, grading tables, flotation tanks and screens can all be used to ensure that the quality standards are met and an even line of high quality spice is obtained.

Packaging

Processed plant materials should be packaged as quickly as possible to prevent deterioration of the product and as a protection against exposure to pest attacks and other sources of contamination. Continuous in-process quality control measures should be implemented to eliminate substandard materials, contaminants and foreign matter prior to and during the final stages of packaging. Processed plant materials should be packaged in clean, dry boxes, sacks, bags or other containers in accordance with standard operating procedures and national and/or regional regulations of the producer and the end-user countries. Materials used for packaging should be non-polluting, clean, dry and in undamaged condition and should conform to the quality requirements for the plant materials concerned. Fragile plant materials should be packaged in rigid containers.

Packaging for seeds and fruits

Vanilla, for most of the world trade, has grades of whole and split beans which are subdivided according to length into various sizes between 12 and 26cm and then put into bundles, each containing 70-100 beans and weighing between 150-200g. These are packed into wax-lined cardboard boxes that hold 20-40 bundles. Beans remaining from the grading and sorting, such as short or broken beans, are bulked together and loosely packed into boxes.

Cardamom, when dried, needs to be kept in polypropylene bags to ensure the flavour components are retained and to prevent re-adsorption of moisture, mould growth and reduce insect pests. A sealing machine with a timer will aid efficient bagging.

Cloves when bagged should be stored in cool, clean dry buildings. Sound, properly dried cloves are relatively free from storage pests. Clove oils should be stored in full, airtight containers, and remain sealed until required. Bulk oil is usually transported in 200 litre metal drums.

Nutmeg & mace is graded according to size, and the bigger the mace, the better the premium. After drying, the nutmeg seeds are put into bags and transported to the processing factory for kernel separation and grading. After grading, nutmeg is bagged, labeled and fumigated prior to export. Nutmegs are usually packed in double-layered linen, jute, sisal or woven bags. If other packaging is used, care should be taken to avoid materials that might lead to sweating and mould development. Mace follows a similar process but there is a need for three months curing before bagging and fumigation and top quality mace is packaged in plastic bags.

Chillis (Capsicums, Chilli peppers and Paprika) that have been dried should be stored in airtight containers away from sunlight as air and light accelerates the rate of bleaching. Chillies should be stored in polyethylene bags or packs and can be stored for a year in a cool, dry and dark place.

Packaging for leaves and stems

It is essential that all material is dry to below 10% moisture content. The leaves and stems of dried herbs should be stored in cool, dark and dry areas of low humidity and polyethylene bags or packs used.

Packaging for flowers and buds

Saffron stigma, when dried and graded, should be stored in airtight containers, in a cool dry place out of the light.

Packaging for roots and Rhizomes

Ginger when fresh should be stored in polyethylene bags with 2% ventilation to prevent dehydration and mould development. Ginger rhizome for bulk shipping can be packed in jute sacks, wooden boxes or lined corrugated cardboard boxes. Processed, dry ginger should be packaged in laminated bags that have low oxygen permeability, and stored in a cool dry environment.

Turmeric is mostly traded as whole rhizome and then processed into powder or oleoresin. The colour constituents of turmeric deteriorate with light, and to a lesser extent under heat and oxidative conditions. Ground turmeric should be stored in UV protective packaging and a cool dry environment.

Packaging for bark, wood and resins

Cinnamon and Cassia quills are graded on colour and pressed into cylindrical bales in jute cloth or corrugated cardboard cartons. Cinnamon and cassia, especially if ground, require polypropylene packaging and polyethylene is not recommended as the flavor components diffuse through it.

Storage

There is a need for quality storage both on-farm and off-farm, with cool stores and warehousing facilities linked to post-harvest crop management. Spices deteriorate rapidly in adverse conditions and should be stored in well-prepared and maintained storage facilities. It is essential the moisture level of the spice to be stored is at a safe level prior to storage. This is usually below 10% moisture. The storehouses should be damp-proof, vermin-proof and bird-proof and where possible have controlled ventilation and devices to control humidity and temperature. A dehumidifier fitted to a storage room, by keeping the atmosphere always dry, can eliminate mould and insect attacks. The room should be fumigated before storage, the walls whitewashed regularly and the facility kept dry. Fumigation against pest infestation should be carried out only when necessary, and should be carried out by licensed or trained personnel. Only registered chemicals authorized by the regulatory authorities of the source country and the countries of intended end-use should be used. All fumigation, fumigation chemicals, and dates of application should be documented. When freezing or saturated steam is used for pest control, the humidity of the materials should be checked after treatment. Some specific examples for major spices are described below.

Pepper has a steady loss of aroma compounds that will be faster in poor storage conditions. Polyethylene-lined gunny bags or laminated HDPE (high density polyethylene) are ideal for pepper storage. Care should be taken with pepper storage to avoid aroma spoilage by other spices in close proximity.

Clove quality is affected by prolonged storage which reduces oil content and may also dull the colour. Oil composition and the ratio of constituents of cloves can change in storage, with the rate and extent of change dependent on the production method.

Nutmeg & mace should not be ground until required because of the loss of volatile constituents when in uncontrolled storage. Incorrectly stored nutmeg oil may also undergo significant composition changes if exposed to high ambient temperatures. Unprotected powders and oil can also absorb unpleasant odours and powders should be stored in sealed containers.

Chillis (Capsicums, Chilli peppers and Paprika) fruit, to ensure storage without mould should be close to 10% moisture or below.

Turmeric dried rhizomes are usually chemically treated or fumigated before storage to prevent insect and fungal infestation. Rhizomes stored in a cool chamber have less disease, rot and pest damage and less sprouting, compared to underground pit storage or heaps outdoors.

7

Ripening of Fruits

Ripe

This word is derived from Saxon word '*Ripi*', which means gather or reap. This is the condition of maximum edible quality attained by the fruit following harvest. Only fruit which becomes mature before harvest can become ripe.

Ripening

Ripening involves a series of changes occurring during early stages of senescence of fruits in which structure and composition of unripe fruit is so altered that it becomes acceptable to eat. Ripening is a complex physiological process resulting in softening, colouring, sweetening and increases in aroma compounds so that ripening fruits are ready to eat or process. The associated physiological or biochemical changes are increased rate of respiration and ethylene production, loss of chlorophyll and continued expansion of cells and conversion of complex metabolities into simple molecules.

Senescence

Senescence can be defined as the final phase in the ontogeny of the plant organ during which a series of essentially irreversible events occur which ultimately leads to cellular breakdown and death.

Major biochemical changes occurring during ripening are:

- Seed maturation (in pollinated fruits)
- Color change

- Abscission (detachment of fruits from parent plant)
- A dramatic rise in respiration rate
- Increased ethylene production
- Change in tissue permeability and cellular
- Softening of pulp tissue
- Decrease in starch: sugar ratio
- Production of volatiles (flavor and smell)
- Development of wax on skin.
- Change in composition of pectic substance
- Change in organic acid.
- Change in protein

Process Associated with Fruit Ripening

Unripe Fruit		**Ethylene** →	**Ripe Fruit**	
Physical Condition	**Chemical cause**	**Enzyme Produced**	**Physical Condition**	**Chemical cause**
Green	Chlorophyll	**Hydrolase** →	Red/Yellow	Anthocyanin/ Curotin
Hard	Pectin	**Pectinase** →	Soft	Less Pectin
Sour	Acid	**Kinase** →	Neutral	Neutral
Mealy	Starch	**Amylase** →	Sweet + juicy	Sugar
Odorless	Large orgs	**Hydrolase** →	Odor	Small orgs

Ripening Techniques

1. Ripening with ethephon / Ethrel

Ethephon (2-choloroethyl-phosphonic acid) is commercially available and is registered for pre-harvest use on a variety of crops for controlling developmental processes or inducing ripening. This chemical is approved for post-harvest use on fruits crops for enhancing ripening. For post-harvest treatments, the known quantity of ethephon is diluted in water and fruits are dipped in the solution for a specified period. This substance ensures that there is uniform ripening of fruits. This technique provides a safe and effective method of ripening of fruits compared to the conventional technique of using calcium carbide.

2. Ripening with Ethylene gas

In this technique, the fruits are exposed to low level of ethylene gas (10-100ppm) in an air-tight ripening chamber for 24 to 72 hours so as to induce ripening. The most important thing in this technique is temperature and relative humidity control inside the ripening chamber, which should range between 15-25^{0}C and 90-95% relative humidity, depending upon the fruit type. Several methods are used to provide proper ethylene concentration in the ripening room.

3. Calcium carbide

In India and many other developing economies, the banana and mango are ripened with the use of calcium carbide, which releases acetylene and ethylene on interaction with moisture coming from fruits. This chemical is harmful to human health and its use for ripening of fruits is band in India under section 44-AA of PFA rules 1954.

4. Smoke treatment

Burning and releasing smoke from leaves, twigs or straw will also hasten ripening in mango.

5. 2, 4-D treatment

2, 4 dichlorophenoxy acetic acid is used in ripening of Guava 2,4,5-T (2,4,5-trichlorophenoxy acetic acid) used in sapota.

6. Potassium permanganate

It absorbs ethylene during transport or storage and delay the ripening in many fruit such as mango, banana and apple.

Degreening

Controlled degreening sometimes is carried out on citrus grown in tropics. Many citrus cultivars mature before green colour disappears from peel. Breakdown of chlorophyll and production of a rich orange colour require exposure to low temperature during maturation, and this explains why mature citrus frequently is sold green on markets in humid tropics, where even night temperatures may not drop much below 25°C. The ceiling of room is relatively high, allowing boxes to be stacked at least four boxes high. A false ceiling is added to provide for adequate air.

Degreening is carried out in ripening rooms, with same ethylene concentrations as above. this process takes 2 to 3 days, and it is again necessary to ventilate daily to ensure that carbon dioxide levels do not exceed 1%. The most rapid degreening occurs at temperatures of 25 to 30°C but the best colour (concentration of peel carotenoids) occurs at 15 to 25°C.

Chemicals that delay ripening

1) **Cytokinins and Kinetins:-**Delays chlorophyll degradation and senescence of leafy vegetable.

2) **Gibberellins: -** Post harvest treatments with GA_3 retard ripening of tomato and bananas GA_3 lowers respiratory rate, retards in climacteric fruits and delays the process of colour changes.

3) **Growth retardants:**

   a) **MH: -** Prevents sprouting of onion bulbs and potato tubers. Also delays ripening of mango.

   b) **Alar: -** Reduce the fruit firmness fruit colour development and early maturation. It is applied before harvest. In lettuce it reduces senescence.

   c) **CCC (Cycocel):-** 2 chloroethyl – trimethyl ammonium chloride used in delaying of senescence of vegetables.

4) **Delaying ripening process by skin coating (waxing):**

Edible waxes are coated on fruits which prevent transportation losses and minimize respiration rate. Sugar wax along with emulsifier is melted and then boiling water is poured slowly to melted ingredients and prepares emulsion.

Emulsifying agents: - TEA (Triethylene amine), TBZ – Tribenzoate, fungicides, SOPP (Sodium ortho phenyl phenate) for control of pencillium in oranges (Take the emulsion in large basins and dip the fruit or basket containing fruits for 30 seconds).

Thiourea: - used for control of penicillium and deplodium on oranges.

Factors Affecting Ripening of Fruits and Vegetables

- Respiration
- Transpiration / water loss
- Ripening Ethylene production
- Pathological stresses
- Mechanical stresses
- Gas composition
- Type of commodity
- Temperature stresses.

Ethylene concentration, exposure time and ripening temperature of important fruits

Commodity	Ethylene concentration (ppm)	Ethylene exposure time	Ripening temperature	Predicted storage after treatment
Apple	10 ppm	6 days	25°C	Less than 4 months at 0°C
Banana	100-150 ppm	24-48 hours	14-18°C	Less than 7 days at 14°C
Mango	100 ppm	12-24 hours	15.5-25°C	Less than 7 days at 10-13°C
Tomato (mature green stage)	100 ppm	3-3.5 days to reach breaker stage and 5-16 days, depending on temperature, to reach full red stage	18°-20°C (65°-68°F) with 90-95% RH	7 days after reaching the
Avocado	10-100	12-48 hours	15-18°C	4.4-13°C
Honey dew melon	100-150	18-24 hours	20-25°C	7-10°C
Kiwifruit	10-100	12-24 hours	0-20°C	0.5-0°C
Orange degreening	1-10	24-72 hours	20-22°C	5-9°C
Stone fruit	10-100	12-72 hours	13-25°C	-0.5-0°C

Optimal ripening conditions for fruit ripening

Temperature	18 to 25°C
Relative humidity	90 to 95%
Ethylene concentration	10 to 100 ppm
Duration of treatment	24 to 74 hours depending on fruit type and stage of maturity
Air circulation	Sufficient to ensure distribution of ethylene within ripening room
Ventilation	Require adequate air exchange in order to prevent accumulation of O_2′ which reduces effectiveness of C_2H_4.

8

Ethylene

Ethylene (C_2H_4 also known as ethene) is a gaseous organic compound that is the simplest of the alkenes chemical structures (alkenes contain a carbon-carbon double bond). The hormone effects of ethylene on general plant growth were first noted in 1864 when leakage from gas street lighting systems caused stunting and deformation of nearby plants. In 1901 Neljubow identified the active component of the gas to be ethylene but it was not until 1934 that Gane identified that plants could synthesize ethylene and in 1935 Crocker proposed ethylene to be the hormone responsible for fruit ripening and senescence of vegetative tissues.

Ethylene play important role in many plant during development processes, including seed germination, vegetative growth, leaf abscission, flowering, senescence and fruit ripening. Ethylene also plays a role in response to water stress, chilling and mechanical injury.

Ethylene and Fruit Ripening

Early examples of the human utilization of ethylene to enhance fruit ripening include the ancient Egyptian practice of gashing figs to enhance ripening responses. All fruit produce small quantity of ethylene during ripening but climacteric fruits produce much larger amounts of ethylene than non-climacteric fruits. Internal ethylene concentration of climacteric fruits varies largely but that in non-climacteric fruits experience little change during

The concentration of ethylene treatments, relative humidity and duration of time require for the ripening of various commodities vary. The effectiveness of ethylene in achieving faster and more uniform ripening depends on the type of fruit being treated, its maturity, the temperature and relative humidity of the ripening room, ethylene concentration and duration of exposure to ethylene. In general, optimal ripening conditions for fruits are:

- Temperature: 18°-25°C
- Relative Humidity: 90-95%
- Ethylene Concentration: 10-100 ppm
- Duration of Treatment: 12-72 hours (depending on fruit type and maturity stage)
- Air Circulation: Sufficient to maintain uniform temperature within the ripening room.
- Ventilation: Adequate air exchange to prevent accumulation of CO_2 which reduce the effectiveness of C_2H_4.

Ethylene Biosynthesis

Methionine

↓

SAM (S-adenosyl methionine)

↓

ACC synthesis enzyme

↓

ACC (1 amino cyclopropane 1 carboxylic acid)

↓

EFE (ethylene forming enzyme)

↓

Ethylene

Other Uses of Ethylene

Ethylene is not only uses in fruit ripening but also uses in some other practices that are under:

Uses	Crops
Post-harvest ripening	Mango, banana, tomatoes, etc.
Pre-harvest ripening	Pepper and tomatoes
Fruit removal	Apple, crabapple and olive
Defoliation	Apple and rose
Fruit loosening	Apple, blackberry, cherries and tangerines
Colour development	Apple , cranberry, fig, grape, pepper and pineapple
De-greening (Pre-harvesting)	Tangerines
De-greening (Post-harvesting)	Citrus fruits
Dehiscence	Walnut
Leaf curing	Tobacco
Flower induction	Pineapple
Sex expiration	Cucumber and squash
Flower bud development	Apple

Impacts of Ethylene on Post Harvest Quality of Fruits and Vegetables

- Russet spotting of lettuce (dark brown spotting on the mid-ribs of lettuce leaves).
- Yellowing or loss of green color (for example, in cucumber, broccoli, kale, spinach, etc.).
- Increased toughness in turnips and asparagus spears.
- Increase or decrease sprouting in potatoes.
- Yellowing and abscission (dropping) of leaves in Brassicas.
- Softening, pitting, and development of off-flavor in peppers, summer squash, and watermelons.
- Browning and discoloration in eggplant pulp and seed.
- Discoloration and off-flavor in sweet potatoes.
- Increased ripening and softening of mature green tomatoes.
- Development of bitter taste in carrots and parsnips.

Steps to Reduce Ethylene Exposure during Storage

- Do not store or transport green leafy vegetables in containers holding ripening fruit (apples, pears, mangoes, tomatoes, bananas).
- If possible, use electric powered equipment in storage areas versus gas powered.
- Remove overripe or rotting fruit from storage loads (these produce higher amounts of ethylene).
- Avoid storing ethylene sensitive products with products that produce high levels of ethylene.
- Increase the ventilation rate of the storage area, assuming that the outside air is ethylene free.
- Use ethylene scrubbers ($KMnO_4$, perlite, alumina, silica gel, verculite, activated carbon and celite) in storage areas to remove ethylene in the air.

9

Respiration

It is a most deteriorating biological process of the harvested fruits and vegetables that leads to the oxidative breakdown of the complex material (carbohydrates or acids) of cell into simpler molecules (CO2 & water) with production of energy. Respiration includes glycolysis metabolism, aerobic metabolism, anaerobic metabolism, trycarboxilic acid cycle and electron transport metabolism processes.

$$C_6H_{12}O_6 + 6O_2 \text{ -----}\rightarrow 6CO_2 + 6H_2O + \text{Energy(686kCal)}$$

(Organic compounds)

In general, the storage life of commodities varies inversely with the rate of respiration. This is because respiration supplies compounds that determine the rate of metabolic processes directly related to quality parameters, eg., firmness, sugar content, aroma, flavor, etc. Commodities and cultivars with higher rates of respiration tend to have shorter storage-life than those with low rates of respiration. Storage life of broccoli, lettuce, peas, spinach, and sweet corn (which have high respiration rates) is short in comparison to apples, cranberries, limes, onions, and potatoes (which have low respiration rates).

Respiration rates of a range of perishable commodities

Class	Range at 5 °C ($mg\ CO_2\ kg^{-1}\ h^{-1}$)	Commodities
Very Low	< 5	Nuts, dates
Low	5 to 10	Apple, citrus, grape, kiwifruit, onion, potato
Moderate	10 to 20	Apricot, banana, cherry, peach, nectarine, pear, plum, fig, cabbage, carrot, lettuce, pepper, tomato
High	20 to 40	Strawberry, blackberry, raspberry, cauliflower, lima bean, avocado
Very High	40 to 60	Artichoke, snap bean, Brussels sprouts, cut flowers
Extremely High	> 60	Asparagus, broccoli, mushroom, pea, spinach, sweet corn

Respiration rates vary among and within commodities. Storage organs such as nuts and tubers have low respiration rates. Tissues with vegetative or floral meristems such as asparagus and broccoli have very high respiration rates. As plant organs mature, their rate of respiration typically declines. This means that commodities harvested during active growth, such as many vegetables and immature fruits, have high respiration rates. Mature fruits, dormant buds and storage organs have relatively low rates.

After harvest, the respiration rate typically declines; slowly in non-climacteric fruits and storage organs, rapidly in vegetative tissues and immature fruits. The rapid decline presumably reflects depletion of respirable substrates that are typically low in such tissues. An important exception to the general decline in respiration following harvest is the rapid and sometimes dramatic rise in respiration during the ripening of climacteric fruit. This rise, which has normally, consists of four distinct phases:

1) Pre-climacteric minimum,
2) Climacteric rise,
3) Climacteric peak, and
4) Post-climacteric decline

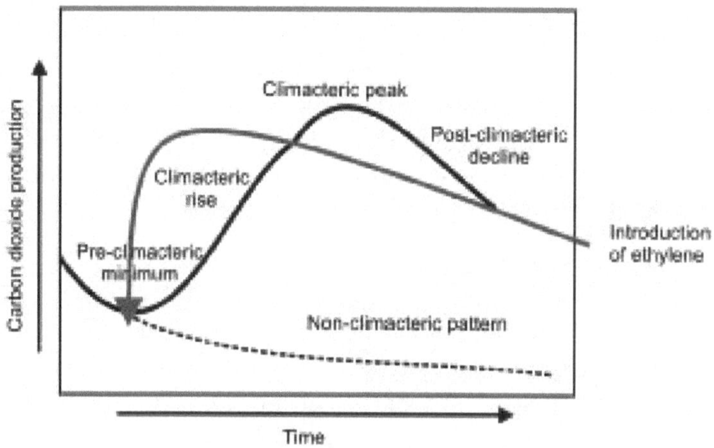

The division of fruits into climacteric and non-climacteric types has been very useful for postharvest physiologists. However, some fruits, for example kiwifruit and cucumber, appear to blur the distinction between the groups. Respiratory rises also occur during stress and other developmental stages, but a true climacteric only occurs coincident with fruit ripening. Following is a general classification of fruits according to their respiratory behavior during ripening. The division of fruits into climacteric and non-climacteric types has been very useful for postharvest physiologists. However, some fruits, for example kiwifruit and cucumber, appear to blur the distinction between the groups. Respiratory rises also occur during stress and other developmental stages, but a true climacteric only occurs coincident with fruit ripening. Following is a general classification of fruits according to their respiratory behavior during ripening:

Climacteric Fruits	Non-Climacteric Fruits
Apple, Apricot, Avocado, Banana, Breadfruit, Cherimoya, Feijoa, Fig, Guava, Jackfruit, Kiwifruit, Mango, Muskmelon, Nectarine, Papaya, Passion fruit, Peach, Pear, Persimmon, Plum, Sapote, Soursop, Tomato, Watermelon, etc.	Blueberry, Cacao, Caju, Cherry, Cucumber, Grape, Grapefruit, Lemon, Lime, Litchi, Olive, Orange, Pepper, Pineapple, Strawberry, Tamarillo, etc.

Respiratoy Quotient (RQ)

The composition of a commodity frequently determines which substrates are utilized in respiration and consequently the value of the respiratory quotient (RQ). The RQ is defined as the ratio of CO_2 produced to O_2 consumed; CO_2 and O_2 can be measured in moles or volumes. Depending on the substrate being oxidized, RQ values for fresh commodities range from 0.7 to 1.3 for aerobic respiration. When carbohydrates are being aerobically respired, the RQ is near 1, while it is < 1 for lipids, and > 1 for organic acids. Very high RQ values usually indicate anaerobic respiration in those tissues that produce ethanol. In such tissues, a rapid change in the RQ can be used as an indication of the shift from aerobic to anaerobic respiration.

Factors Affecting Respiration

Respiration is affected by a wide range of environmental factors that include light, chemical stress (eg., fumigants), radiation stress, water stress, growth regulators, and pathogen attack. The most important postharvest factors are temperature, atmospheric composition and physical stress.

Temperature

Without a doubt, the most important factor affecting postharvest life is temperature. This is because temperature has a profound effect on the rates of biological reactions, eg., metabolism and respiration. Over the physiological range of most crops, ie., 0 to 30 °C (32 to 86 °F), increased temperatures cause an exponential rise in respiration. The Van't Hoff Rule states that the velocity of a biological reaction increases 2 to 3-fold for every 10 °C (18 °F) rise in temperature.

Atmospheric Composition

Adequate O_2 levels are required to maintain aerobic respiration. The exact level of O_2 that reduces respiration while still permitting aerobic respiration varies with commodity. In most crops, O_2 level around 2 to 3% produces a beneficial reduction in the rate of respiration and other metabolic reactions. Levels as low as 1% improve the storage life of some crops, eg., apples, but only when the storage temperature is optimal. At higher storage temperatures, the demand for ATP may outstrip the supply and promote anaerobic.

The need for adequate O_2 should be considered in selecting the various postharvest handling procedures, such as waxing and other surface coatings, film wrapping and packaging

Increasing the CO_2 level around some commodities reduces respiration, delays senescence and retards fungal growth.

Physical Stress

Even mild physical stress can perturb respiration, while physical abuse can cause a substantial rise in respiration that is often associated with increased ethylene evolution. The signal produced by physical stress migrates from the site of injury and induces a wide range of physiological changes in adjacent, non-wounded tissue.

10

Storage

Proper marketing of perishable commodities such as fruit and vegetables often requires some storage to balance day-to-day fluctuation between harvest and sale or for long term storage. Storage of horticulture produce is attempeted with a view to extend period of availability. Storage controls shelf life of produce by controlling rate of respiration, transpiration, ripening and biochemical changes all of which are responsible for shelf decomposition/deterioration of produce. Further, different types of storage have different attributes in minimizing microbial infection and thus add to better storability of produce.

Objective of Storage

- Slow the biological activity and micro-organism growth.
- Avoid glut and distress sale in the market, thus prolonging the market period.
- In long-term storage, making the food available in off-season.
- Regulate the market in an orderly manner.
- Reduce the transpirational, respiration, and ethylene production rate.

Tips for Storage of High Quality Horticultural Produce

- Store only high quality produce, free of damage, decay and of proper maturity (not over-ripe or under-mature).
- Knowledge of the appropriate storage conditions.

- Avoid lower than recommended temperatures in storage, because many commodities are susceptible to damage from freezing or chilling.
- Do not over load storage rooms or stack containers closely.
- Provide adequate ventilation in the storage room.
- Keep storage rooms clean.
- Storage facilities should be protected from rodents by keeping the immediate outdoor area clean, and free from trash and weeds.
- Containers must be well ventilated and strong enough to with stand stacking. Do not stack containers beyond their stacking strength.
- Avoid storing ethylene sensitive commodities with those that produce ethylene.
- Avoid storing produce known for emitting strong odors (apples, garlic, onions, turnips, cabbages and potatoes) with odor-absorbing commodities.
- Inspect stored produce regularly for signs of injury, water loss, damage and disease.
- Remove damaged or diseased produce to prevent the spread of problems.

Factors Affecting of Storage

1. Temperature
2. Relative Humidity
3. Ventilation
4. Air Velocity
5. Atmospheric gases composition
6. Light
7. Commodities
8. Post Harvest Treatment

Storage Method

1. Traditional Storage
2. Advanced Storage

Traditional Storage

a) Pit Storage
b) Clamp Storage
c) Barns Storage
d) Cellars Storage
e) Sand and Coir Storage
f) High Altitude Storage
g) Zero Energy Cool Chamber

Zero Energy Cool Chamber

The zero energy cool chamber used to enhance the shelf life of fresh fruit and vegetables by maintaining the temperature and relative humidity during summer and winter suasion naturally with the application of water. It is based on the principle of direct evaporation cooling and hence, does not require any electricity or power to operate, and all the material (Brick, sand, bamboo, khaskhas, etc.) required to make the cool chamber are available easily and are reusable. For zero energy cool chamber construction upland is selected where there is a lot of aeration and should be near to the source of water supply

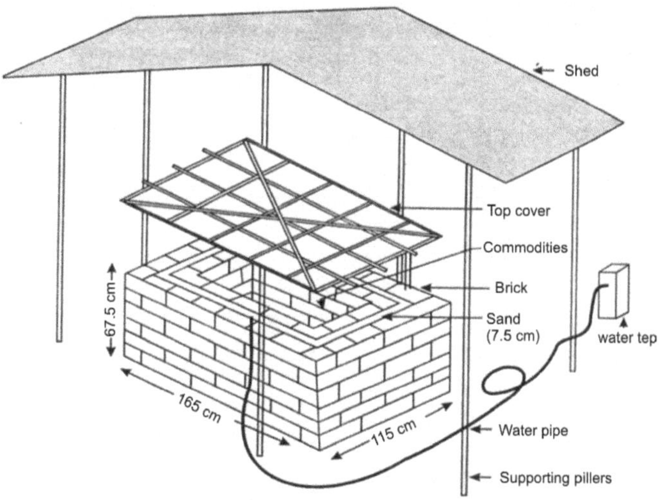

Fig 1 : Zero Energy Cool Chamber

and floor (165 cm x115cm) is made by bricks. A double wall is erected upto a height of 67.5 cm and leaving a cavity with approximately 7.5 cm space between the inner and outer brick walls. The cavity (7.5 cm) between the double walls is filled with sand and drenched with water. A lid (165cmx115cm) is made using straw/ dry grass on bamboo frame. The shed is made over the chamber in order to avoid direct sun or rain.

The whole structure is made wet by sprinkling water daily twice (morning and evening) in order to achieve desired temperature and relative humidity or fix a drip system with plastic pipes and micro tubes connected to an overhead water source. The horticultural produce is kept in perforated plastic crates covered with a thin polyethylene sheet. Maximum and minimum thermometer and hygrometer are placed in the chamber to note temperature and relative humidity, respectively

Shelf life of fruits and vegetables in zero energy chamber (days)

Produce	Time of storage	Outside	Cool chamber
Leaf vegetables	Summer	<1	3
Leaf vegetables	Winter	3	8-10
Potato	Spring/Summer	40	97
Mango	Summer	4	8
Orange	Winter	8-10	20-30
Tomato	Summer	7	15
Pointed gourd	Summer	2	5
Bitter gourd	Summer	2	6
Okra	Summer	1	6
Carrot	Spring	5	12
Cauliflower	Spring	7	12
Green chilli	Summer	3	6

Advanced Storage

A. Cold storage

It maintains low temperature and humidity throughout the storage period. Temperature which is just sufficient to lower down metabolic activity of produce is maintained. Care is taken that the inside temperature may not cause chilling injury to the produce. Depending upon type of produce different temperature and humidity are maintained under cold storage.

Recommended temperature and relative humidity conditions for fruits storage

Name of commodity	Temp (°C)	RH (%)	Approximate Shelf-Life
Acid lime,	8-10		6-8 weeks
Apple	-1-0	90-95	4-8 months
Apricot	0-1	90-95	1-2 weeks
Asian pear	0-1	90-95	2 months
Avocado	8-12	85-90	1.5 weeks
Banana	13	90-95	1-4 weeks
Ber	3	90-95	4 weeks
Fig	0-1.7	85-90	7weeks
Date	6.7	85-90	2 weeks
Grape	0-1	90-95	2-8 weeks
Guava	8-10	90-95	2-3 weeks
Kiwi, Chinese gooseberry	0	90-95	3-5 months
Lemon	10-13	85-90	1-6 months
Loquat	11	90	3 weeks
Lychee, Litchi	5.7	90-95	3-5 weeks
Mandarin (Kinnow)	8-10	90-95	2 months
Mango	8-9	85-90	2-4 weeks
Mangosteen	4-6		2-3 weeks
Nectarine	-0.5-0	90-95	2-4 weeks
Papaya	9-10	90-95	1-3 weeks
Passion fruit	5.6-7.2	85-90	3 weeks
Peach	0-0.3	90-95	2-4 weeks
Pear	-1.5-0.5	90-95	4-5 months
Persimmon	0-1.7	85-90	7 weeks
Pineapple	7-13	85-90	2-4 weeks
Plum and prunes	0-1	90-95	2-5 weeks
Pomegrante	5	90-95	2 months
Strawberry	0	90-95	7-10 weeks
Sweet cherries	-1-0.5	90-95	2-3 weeks
Sweet Orange	2-5		2-3 month

Recommended temperature and relative humidity conditions for vegetables storage

Name of commodity	Temp (°C)	RH (%)	Approximate Shelf-Life
Asparagus, green	1-2	95-100	2-3 weeks
Beans	4-7	90-95	7-10 days
Bell Pepper	7-10	90-95	2-3 weeks
Bitter gourd	10-12	85-90	2-3 weeks
Brinjal	10-11	92	3-4 week
Broccoli	0	95-100	10-14 days
Cabbage	0	90-95	3-6 weeks
Carrots	0.4-4	93-99	6-8 months
Cauliflower	0	90-95	3-4 weeks
Eggplant	10-12	90-95	1-2 weeks
Garlic	0	65-70	6-7 months
Ginger	13	65-70	6 months
Lettuces	0	90-95	2-3 weeks
Mushrooms	0	90	7-14 days
Okra	8.9	90-95	7-10 days
Onion	0	65-70	20-24 weeks
Peas	0	90-95	1-2 weeks
Potato	3-4.4	85	34 weeks
Radish	0	90-95	1-2 months
Tomato (Ripe)(Unripe)	7.2 8.9-10	90 85-90	1weeks 4-5 weeks
Turnip	0	90-95	4-5 months
Watermelon	10-15	90-95	2-3 weeks

Hypobaric Storage

In such type of storage, the produce is storage under low pressure atmosphere, vacuum tight and refrigerated condition. The pressure of 1.3 to 13 kPa is maintained under storage. The due to lowering of pressure under storage, the pressure of individual gases are lowered. The CO_2 and other volatiles like ethylene produced during respiration in course of storage are removed and controlled level of O_2 is maintained. Many time flavour of the produce is affected.

Controlled Atmosphere (CA) Storage

Storage of commodities under gaseous atmosphere condition, like gas storage, where the composition of atmosphere is controlled accurately. Thus the levels of carbon dioxide, oxygen, nitrogen, ethylene, and metabolic volatiles in the atmosphere may be manipulated. Controlled atmosphere storage generally refers to keeping produce at decreased oxygen and increased carbon dioxide concentrations and at suitable range of temperature and RH. The modified atmosphere and controlled atmosphere differ only in degree of control and CA is more exact. This storage method in combination with refrigeration markedly enhances storage life of fruit and vegetables.

Benefits of CA Storage

- Slow down respiration and ethylene production rates, softening and retard senescence of horticultural produce.
- Reduce fruit sensitivity to ethylene action
- Alleviate certain physiological disorders such as chilling injury of various commodities, russet spotting in lettuce and some storage disorders including, scald of apples.

Harmful Effects of CA Storage

- Initiation or aggravation of certain physiological disorders can occur, such as blackheart in potatoes, brown stain on lettuce and brown heart in apples and pears.
- Irregular ripening of fruits, such as banana, mango, pear and tomato, can result from exposure to O_2 levels below 2% or CO_2 levels above 5% for more than 2 to 4 weeks.
- Off- flavors and off-odours at very low O_2 or very high CO_2 concentration may develop as a result of anaerobic respiration and fermentative metabolism.

Modified Atmospheric Storage (MA)

Storage in which inside atmosphere is modified in such a way that its composition is other than that of air constitutes MA storage. The modification is achieved by manipulating O_2, CO_2 and N_2 is storage atmosphere. It favoures extension in self life of produce owing to reduced respiration and ethylene production, minimize water loss and nutrient decomposition, reduced microbial growth and spoilage and inhibition of ripening and senescence.

Recommended CA or MA conditions for important fruits

Commodity	Temperature (°C)	% O_2	% CO_2
Apple	0-5	1-2	0-3
Apricot	0-5	2-3	2-3
Avocado	5-13	2-5	2-5
Banana*	12-16	2-5	2-5
Cherry, sweet	0-5	3-10	10-15
Fig	0-5	5	15
Grape	0-5	0	0
Grapefruit	10-15	3-10	5-10
Kiwifruit	0-5	2	5
Lemon	10-15	5	0-5
Lime	10-15	5	0-10
Mango*	10-15	3-7	5-8
Nectarine	0-5	1-2	5
Olive	8-12	2-5	5-10
Orange	5-10	10	5
Papaya	10-15	5	10
Peach, clingstone	0-5	1-2	3-5
Pear, European	0-5	1-3	0-3
Pineapple	10-15	5	10
Strawberry	0-5	10	15-20
Nuts & dried fruits	0-25	0-1	0-100

* CA is especially beneficial during transit

Recommended CA or MA conditions for important vegetables

Commodity	Temperature (°C)	% O_2	% CO_2
Artichokes	0-5	2-3	3-5
Asparagus	1-5	Air	10-14
Beans, green	5-10	2-3	4-7
Beet	0-5	0	0
Broccoli	0-5	1-2	5-10
Brussels sprouts	0-5	1-2	5-7
Cabbage	0-5	2-3	3-6
Cantaloupes	2-7	3-5	10-20
Carrot	0-5	0	0

Contd.

Cauliflower	0-5	2-3	3-4
Celery	0-5	2-4	0
Corn (sweet)	0-5	2-4	10-20
Cucumber	8-12	3-5	0
Leek	0-5	1-2	3-5
Lettuce	0-5	2-5	0
Mushroom	0-5	Air	10-15
Okra	7-12	Air	4-10
Onions (bulb)	0-5	1-2	0-10
Pepper (bell)	5-12	2-5	2-5
Pepper, chilli	8-12	3-5	0
Potato	4-12	0	0
Radish (topped)	0-5	1-2	2-3
Spinach	0-5	Air	10-20
Tomatoes (green) ripe	12-20 10-15	3-5 3-5	3-5 3-5

11

Packaging

Packaging is an important consideration in vegetable and fruit market. The use of properly designed containers for transporting and marketing of vegetables is significantly reduce losses and maintain their freshness succulence and quality for longer period. Packaging also provides protection from mechanical damage and undesirable physiological changes and pathological deterioration during storage, transportation and marketing.

Many vegetables are transported in gunny bags and bamboo baskets. Packaging material such as polythene films, paper boars, and boxes lived with polythene and other materials can effectively prolong the shelf life of vegetables. By using plastic films vegetables can be protected from dry air. Polythene packaging provides modified atmosphere and consequently reduces decay, softening, and loss of solids. The thickness and permeability to CO_2, O_2 and water vapour of films needs to be standardized for each vegetable.

The Function of Packaging or Why Package Produce?

A significant percentage of produce buyer and consumer complaints may be traced to container failure because of poor design or inappropriate selection and use. A properly designed produce container should contain, protect, and identify the produce, satisfying everyone from grower to consumer.

The main objective of packaging is to keep the fruits, vegetable and root crops in good condition until it is sold and consumed.

Characteristics of packaging

The characteristics of packaging are to contain, to protect, to communicate and to market the product.

A. To contain produce

- As an efficient handling unit, easy to be handled by one person.
- As a marketable unit. e.g. units with the same content and weight.

B. To protect produce against

- Rough handling during loading, unloading and transport - rigid crate.
- Pressure during stacking.
- Moisture or water loss with consequent weight and appearance loss.
- Heat: air flow through crate or box via ventilation holes.
- Fumigation possible through ventilation holes.

C. To communicate

- Identification: a label with country of origin, volume, type or variety of product, manufacturing and expiry date, etc. printed on it.
- Marketing and Advertising: recognizable trade name and trademark.

D. To market the product

- Proper packaging will lead to reduced injuries of fruits and vegetables and subsequently to improvement of appearance.
- Standard units (weight, count) of a certain produce will increase speed and efficiency of marketing.
- With reduced costs of transport and handling, stacking and combining of packages into layer units like pallets is possible. A more efficient use of space and reduced losses will lower the marketing costs.
- Labels and slots facilitate inspection.

Type of Packaging

1) **Plastic film bags: -** Widely used for consumer size packs in fruit and vegetables marketing. Retain water vapour so as to reduce H_2O loss from the content.
2) **Plastic boxes: -** They are rigid containers most suited for packaging soft and delicate commodities.
3) **Net / mesh bags: -** Widely used for packing fruits like apple, citrus, guava, sapota *etc.*
4) **Sleeve packs: -** Immobilization of packed fruits, superior visibility that gives a good sales appeal.
5) **Cling film: -** Ideal packaging for low water vapour transmission rate, high gas permeability.
6) **Shrink film or stretch film: -** Among the specialised plastic packaging systems, shrink packaging or commonly known as shrink wrapping and stretch wrapping are very common and widely used. They are considered to be totally different in terms of material and operation.

In the case of shrink wrapping, shrink film is used as the basic material and heat forms an important part of the operation, whereas, in the case of stretch wrapping, stretch film is used as the basic material and no heat is applied during the operation. Shrink Wrapping is done in 3 or 4 stages, namely:

- Wrapping (sleeve wrapping or over-wrapping)
- Sealing (necessary only for over-wrapping)
- Shrinking (with application of hot air), and
- Cooling

Stretch wrapping is done only in two stages, namely, wrapping and sealing (most of the time even without a sealer).

7) **Antimicrobial packaging: -** Incorporating antimicrobial agents into polymer surface coating sand surface attachments.
8) **Wooden packaging: -** used for packing fruits and vegetable.

The advantages of wooden crates are:

- The crates can be manufactured and repaired locally.
- Wood is relatively resistant to different weather conditions and (sea) water.

- Wooden crates are often used on more than one journey and have a higher efficiency for larger fruits, e.g. watermelons.
- Most crates have good ventilation and fast pre-cooling is possible.

Disadvantages of wooden crates are:

- Untreated wood can easily become contaminated with fungi and bacteria.
- Treatment of wooden crates with paint or other chemicals may cause produce deterioration.
- The material may be too hard or rough for produce like soft fruits, and therefore liners of a soft material may be needed.
- Disposal of the crates after use.
- Manufacturing of wooden crates puts an extra claim on the natural forest resources.

9) **Modified atmosphere packaging :-** It is the packaging of perishable products. Modified atmosphere packaging (MAP) of fresh fruits and vegetables is based on modifying the levels of O_2 and CO_2 in the atmosphere produced inside a package sealed with some type of polymer film. It is desirable that the natural interaction that occurs between the respiration of the product and the packaging generates an atmosphere with low levels of O_2 and / or a high concentration of CO_2. The growth of organisms that cause decay is thereby reduced and the life of the product is extended. Additionally, the desired atmosphere can reduce the respiration rate, and ethylene production, physiological changes. For example, it can inhibit chemical, enzymatic and microbiological mechanisms associated with the decay of fresh products, thus avoiding the use of other chemical or thermal process such as freezing, dehydration, and sterilization. Modified atmosphere packaging is two types:-

Passive Modified Atmosphere Packaging

Modified atmospheres can be obtained passively between plant material and sealed package or intentionally using determined concentrations of gases. Modified atmosphere is formed as a result of vegetable respiration, which consumes CO_2 and releases O_2 in sealed package. In passive modification, the respiring product is placed in a polymeric package and sealed hermetically. Only the respiration of the product and the gas permeability of the film influence the change in gaseous composition of the environment surrounding the product. If the product's respiration characteristics are properly matched to the film permeability values, then a beneficial modified atmosphere can be passively created within a package. The polymer itself variably restricts gas exchange between the internal and external environments due to its selective permeability to O_2 and CO_2. After a period of time, the Modified Atmosphere Packaging for Perishable Plant Products system reaches an equilibrium atmosphere containing of lower concentrations of O_2 and higher concentrations of CO_2 than in atmospheric air.

Active Modified Atmosphere Packaging

The concept of active packaging has been developed to adjust the deficiencies in passive packaging such as when a film is a good barrier to moisture, but not to oxygen, the film can still be used along with an oxygen scavenger to exclude oxygen from the pack. An intentionally or actively obtained modified atmosphere occurs when the desired gas mixture is introduced into the container before sealing. In this way, atmospheric balance inside the package is reached faster or almost immediately. Sometimes, certain additives are incorporated into the polymeric packaging film or within packaging containers to modify the headspace atmosphere and to extend shelf-life. Another process is the acceleration of atmospheric balance under partial vacuum packaging is the process of removing the air before sealing, reducing the free space. Although the active modification of the atmosphere within the package incurs additional costs, the advantage is that the desired atmosphere is securely achieved in considerably less time.

10) **Vacuum packaging: –** Where as MAP and CAP mostly operate in ambient pressure (101 kpa), storage at reduced atmosphere has been experimented and is known as vaccum packaging. Packaging the products in film of low oxygen permeability and sealing it after evacuating the air. Apple can be stored well in below pressure of 10 kpa.

11) **Bamboo mat holed boxes: -**Suitable for transportation of apple.

12) **Polypropelene boxes: -** Highly suitable for long markets it can be reused.

13) **Corrugated fibre board: -**Suitable for fruit and vegetable and most economical.

Type of Packaging Materials

1. Sacks: flexible, made of plastic or jute.
   i. Bags: small size sack
   ii. Nets: sacks made of open mesh
2. Wooden crates.
3. Carton or fiber-board boxes.
4. Plastic crates.
5. Pallet boxes and shipping containers.
6. Baskets: made of woven strips of leaves, bamboo, plastic, etc.

12

Physiological Disorders and Post Harvest Disease of Horticulture Crops

Physiological disorders refer to non-parasitic or inanimate disease of fruit crops, or it is the breakdown of tissue that is not caused by either invasion by pathogens or by mechanical damage. They may develop in response to an adverse environment, especially temperature or to a nutritional deficiency during growth and development. Physiological disorder is broadly classified into two broad categories. They are:

- Due to unfavorable environmental condition.
- Nutritional deficiency.

The productivity as well as the quality of fruit crops is affected to a greater extent due to the physiological and nutritional disorders. Disturbance in the plant metabolic activities resulting from an excess or deficit of environmental variables like temperature, light, aeration and nutritional imbalances result in crop disorders. In fruit crops, the deficiency of micronutrients causes many more disorders than that of macronutrients. Nutritional disorders have become widespread with diminishing use of organic manures, adoption of high density planting, use of root stocks for dwarfing, disease and salt tolerance, unbalanced NPK fertilizer application and extension of horticulture to marginal lands. To get high quality fruit and yields, micronutrient deficiencies have to be detected before visual symptoms are expressed.

Physiological disorder of fruits, vegetables and flowers

Disorder	Fruits	Cause
Bitter Pit	Apple	Ca Deficiency
Black Heart	Apple	Low Temperature
Lentical Blotch	Apple	Ca & Water Deficiency
Cork Spot	Apple	Ca Deficiency
Collar Rot	Apple	Low Temperature
Lentical break down	Apple	Ca Deficiency
Cracking	Apple	Ca Deficiency
Internal Cork	Apple	B Deficiency
Internal break down	Apple	Ca Deficiency
Jonathan Spot	Apple	Ca & Water Deficiency
Water Core	Apple	Ca & B Deficiency, Excess of Water and High Temperature
Fruit Necrosis	Aonla	B Deficiency
Pit Burning	Apricot	High Temperature
End Spot	Avocado	Ca Deficiency
Improper Finger Filling	Banana	K Deficiency
Hypocotyl Necrosis	Bean	Ca Deficiency
Internal Browning	Brussel Sprout	Ca Deficiency
Internal Tip Burn	Cabbage	Ca Deficiency
Cavity Spot	Carrot	Ca Deficiency
Cracking	Carrot	Ca Deficiency
Crown Choking	Coconut	B Deficiency
Fruit Mummification	Custard Apple	Water Deficiency
Fruit Cracking	Citrus and Litchi	Excess of Water
Milarandage and Hen and Chicken Disorder	Grape	B Deficiency
Tip Burn	Lettuce	Ca Deficiency
Soft Nose	Mango	Ca Deficiency
Internal Necrosis	Mango	B Deficiency
Cavity spot	Parsnip	Ca Deficiency
Blackening	Peaches	High Temperature
Cork Spot	Pear	Ca Deficiency
Black End	Pear	Water Deficiency
Side Cracking	Prunes	High Temperature
Fruit Cracking	Pomegranate	B Deficiency

Contd.

Sprout Failure	Potato	Ca Deficiency
Tip Burn	Potato	Ca Deficiency
Leaf Tip Burn	Strawberry	Ca Deficiency
Blossom End Rot , Black Seed and Cracking	Tomato	Ca Deficiency
Blossom End Rot	Water Melon	Ca Deficiency
Vascular Plugging	Rose	Insufficient water absorption
Blacking of petals	Rose	Low temperature and high anthocyanin content
Bull head	Rose	Abnormal production of cytokinins and gibberellins
Limp neck	Rose	
Calyx splitting	Carnation	More temperature fluctuation and excessive nitrogen
Sleepiness	Carnation	Exposure to ethylene, water stress and high temperature
Quilling of florets	Chrysanthemum	Low temperature
Petal burn	Chrysanthemum	High temperature
Blindness	Gladiolus	Unfavorable climate condition
Topple and bud rot	Gladiolus	Ca Deficiency
Negative geotropism	Gladiolus	Uneven distribution of auxins and transport vertically
Dry sepal injury	Orchids	High humidity and smog
Deformation flowers and browning of throat	Orchids	Low temperature and chilling injury

Post Harvest Disease

Losses due to postharvest disease are affected by a great number of factors including:

- Commodity type
- Cultivar susceptibility to postharvest disease
- The postharvest environment (temperature, relative humidity, atmosphere composition, etc.)
- Produce maturity and ripeness stage
- Treatments used for disease control
- Produce handling methods
- Postharvest hygiene.

Post harvest disease of fruits and its pathogen

Disease	Fruit	Pathogen	
		Anamorph	Teleomorph
Blue mould	Pome Fruit	*Pemcillium* spp.	
Grey mould	Pome Fruit	*Botrytis ctnerea*	*Botryotinia fuckeliana*
Bitter rot	Pome Fruit	*Colletotrtchum gloeosporioides*	*Glomerella cingulata*
Alternaria rot	Pome Fruit	*Alternaria* spp.	
Mucor rot	Pome Fruit		*Mucor piriformts*
Brown rot	Stone Fruit	*Monilia* spp.	*Monilinia fiucticola (syn. Sclerotinia fructicola)*
Rhizopus rot	Stone Fruit		*Rhizopus spp.(mostly R.stolonifer)*
Grey mould	Stone Fruit	*Botrytis ctnerea*	*Botryotinia fuckeliana*
Blue mould	Stone Fruit	*Penicillium* spp.	
Alternaria rot	Stone Fruit	*Alternaria alternata*	
Grey mould	Grape	*Botrgtis ctnerea*	*Botryotiniafuckeltana*
Blue mould	Grape	*Penrrctllium* spp.	
Rhizopus rot	Grape		*Rhrzopus spp.*
Grey mould	Berries	*Botrgtis cinerea*	*Botryotinia fuckeliana*
Rhizopus rot	Berries		*Rhrzopus spp*
Cladosporium rot	Berries	*Cladosporium* spp.	
Blue mould	Berries	*Penicillium* spp.	
Blue mould	Citrus Fruit	*Penicillium italicum*	
Green mould	Citrus Fruit	*Penicillium digitatum*	
Black centre rot	Citrus Fruit	*Alternaria citrt*	
Stem end rot	Citrus Fruit	*Phomopsis citri*	
Brown rot	Citrus Fruit		*Diaporthe citrt, Phytophthora citrophthora, and P. parasitica*
Anthracnose	Avocado	*Colletotrichum Gloeosporioides and C. acutatum*	*Glomerella cingulata*
Stem end rot	Avocado	*Dothiorella* spp., *Lasiodiplodia theobromae and Stilbella cinnabarina Phomopsis perseae*	*Botryosphaeria* spp. *andThyronectria pseudotrichia*

Contd.

Bacterial soft rot	Avocado	*Erwinia carotovora*	
Anthracnose		*Colletotrichum musae*	
Crown rot	Banana	*Various fungi including Fusarium* spp., *Verticillium* spp., *Acremonium sp. and Colletotrichum musae*	
Black end	Banana	*Various fungi including Colletotrichum musae, Nigrospora sphaericaand Fusarium* spp.	
Ceratocystisfruit rot	Banana	*Thielaviopsis paradoxa*	*Ceratocystis paradoxa*
Anthracnose	Mango	*Colletotrichum gloeosporioides and C. acutatum*	*Glomerella cingulata*
Stem end rot	Mango	*Dothiorella* spp., *Lasiodiplodia theobromae, Phomopsis mangiferae and Pestalotiopsis mangiferae*	*Botrgosphaena* spp.
Rhizopus rot	Mango		*Rhizopus stolonifer*
Black mould	Mango	*Aspergillus niger*	
Alternaria rot	Mango	*Alternaria alternata*	
Grey mould	Mango	*Botrytis cinerea*	*Botryotinia fuckeliana*
Blue mould	Mango	*Penicillium expansum*	
Mucor rot	Mango		*Mucor circinelloides*
Anthracnose	Papaya	*Colletotrichum* spp.	
Black rot	Papaya	*Phoma caricae-papayae*	*Mycosphaerella caricae*
Phomopsis rot	Papaya	*Phomopsis caricae -papayae*	
Rhizopus rot	Papaya		*Rhizopus stolonifer*
Phytophthora fruit rot	Papaya		*Phytophthora palmivora*
Water blister		*Thielaviopsis paradoxa*	*Ceratocystis paradoxa*
Fruitlet core rot		*Penicillium funiculosum andFusarium moniliforme var subglutinans*	*Gibberella fujikuroi var subglutinans*
Yeasty rot		*Saccharomyces* spp.	
Bacterial brown rot		*Erwinia ananas*	

Post harvest disease of vegetables and its pathogen

Disease	Vegetable	Pathogen	
		Anamorph	Teleomorph
Bacterial soft rots	Cucurbits	*Erwinia* spp., *Bacillus polymyxa, Pseudomonas syringae andXanthomonas campestris*	
Grey mould	Cucurbits	*Botrytis cinerea*	*Botryotinia fuckeliana*
Fusarium rot	Cucurbits	*Fusarium* spp.	
Alternaria rot	Cucurbits	*Alternaria* spp	
Charcoal rot	Cucurbits	*Macrophomina phaseolina*	
Cottony leak	Cucurbits		*Pythium spp.*
Rhizopus rot	Cucurbits		*Rhrzopus spp.*
Bacterial soft rots	Tomato, Eggplant and Capsicum	*Erwinia* spp., *Bacillus polymyxa, Pseudomonas* spp. *and Xanthomonas campestris*	
Grey mould	Tomato, Eggplant and Capsicum	*Botrytis cinerea*	*Botryotinia fuckeliana*
Fusarium rot	Tomato, Eggplant and Capsicum	*Fusarium* spp.	
Alternaria rot	Tomato, Eggplant and Capsicum	*Alternaria alternata*	
Cladosporium rot	Tomato, Eggplant and Capsicum	*Ctadosporium* spp.	
Rhizopus rot	Tomato, Eggplant and Capsicum		*Rhizopus* spp.
Watery soft rot	Tomato, Eggplant and Capsicum		*Sclerotinia* spp.
Cottony leak	Tomato, Eggplant and Capsicum		*Pythium* spp.

Contd.

Sclerotium rot	Tomato, Eggplant and Capsicum	*Sclerotium rolfsii*	*Athelia rolfsii*
Grey mould	Legumes	*Botrytis cinerea and B. fabae*	*Botryotinia fuckeliana*
White mould and Watery soft rot	Legumes		*Sclerotinina* spp
Cottony leak	Legumes		*Pythium* spp
Sclerotium rot	Legumes	*Sclerotium rolfsii (sclerotial state)*	*Athelia rolfsii*
Bacterial soft rot	Brassicas	*Erwinia* spp., *Bacillus* spp., *Pseudomonas* spp. *and Xanthomonas campestris*	
Grey mould	Brassicas	*Botrytis cinerea*	*Botryotinia fuckeliana*
Alternaria rot	Brassicas	*Alternarta* spp.	
Watery soft rot	Brassicas	*Sclerotinia* spp.	
Phytophthora rot	Brassicas	*Phytophthora porri*	
Bacterial soft rot	Leafy Vegetables	*Erwinia* spp., *Pseudomonas* spp. *and Xanthomonas campestris*	
Grey mould	Leafy Vegetables	*Botrytis cinerea*	*Botryotinia fuckeliana*
Watery soft rot	Leafy Vegetables		*Sclerotinia* spp.
Bacterial soft rot	Onions	*Erwinia* spp., *Lactobaciiius* spp. *and Pseudomonas* spp.	
Black mould rot	Onions	*Aspergillus niger*	
Fusarium basal rot	Onions	*Fusarium oxysporum sp.cepae*	
Smudge	Onions	*Colletotrichum circinans*	
Bacterial soft rot	Potatoes	*Erwinia* spp.	
Dry rot	Potatoes	*Fusarium* spp.	*Gibberella* spp.
Gangrene	Potatoes	*Phoma exigua var. exigua and var. foveata*	
Black scurf	Potatoes	*Rhizoctonia solani*	*Thanatephorus cucmeris*

Contd.

Silver scurf	Potatoes	*Helminthosporium solani*	
Skin spot	Potatoes	*Polyscytalum pustulans*	
Bacterial soft rot	Carrots	*Erwinia* spp. *and Pseudomonas* spp.	
Rhizopus rot	Carrots		*Rhizopus* spp.
Grey mould	Carrots	*Botrytis cinerea*	
Watery soft rot	Carrots		*Sclerotinia* spp.
Sclerotium rot	Carrots	*Sclerotium rolfsii*	*Athelia rolfsii*
Chalara and Thielaviopsis rots	Carrots	*Chalara thielavioides and Thielaviopsis basicola*	

Postharvest Disease Control and Prevention

- Fungicides
- Maintenance of host resistance to infection through manipulation of the postharvest environment
- Hygiene practices
- Pre-harvest tractors
- Prevention of injury
- Heat treatments
- Lionizing radiation
- Biological control
- Constitutive and induced host resistance

13

Principles and Methods of Preservation of Fruits and Vegetables

Preservation means just protect the foods against the spoilage, but scientifically it may be defined as a science which deals with the process for prevention of decay or spoilage of the food is called preservation.

In other words, just controlling the physical, chemical or microbial changes in the foods is called preservation.

1. **Physical Changes:** Colour, flavour, texture and taste *etc.*
2. **Chemical Changes:** Carbohydrate, fats, proteins, vitamins and minerals.
3. **Microbial Changes:** Mould, yeasts and bacteria

Why to Preserve the Food?

1. To supply to increase the shelf life of the food for increasing the supply.
2. To make the seasonal fruits available throughout the year.
3. To add the variety to the diet.
4. To save time by reducing preparation, time and energy by fire.
5. To stabilize the prices of the food in the market.
6. To improve the health of the population.

Principles of Preservation: There are three main principles:

A. Prevention / delay the microbial decomposition of the food.

B. Prevention / delay the shelf decomposition of the food.

C. Prevention of damage by insects, animals, mechanical causes *etc.*

A. Prevention / delay the microbial decomposition of the food

1. By keeping out the micro organisms -Asepsis
2. By Removal of micro organisms -Filtration
3. By Hindering the growth and activity of micro organisms - Anaerobic condition
4. By killing the micro organisms -Exposing at high temperature

1. Asepsis: It means preventing the entry of micro organisms by maintaining of general cleanliness, while picking, grading, packing and transporting of fruits and vegetables, increase their keeping quality and the product prepared from them will be superior quality.

2. Filtration: Fruits juice, bear, soft drinks, wines *etc.* enter through bacteria proof filters which are made of Asbestos pad or unglazed porcelain type of materials. These filters contain the micro organisms and allow the water or juice to percolate though with or without pressure.

3. Anaerobic conditions: It can be maintained by:

(i) Replacing the O_2 by CO_2 carbonation

(ii) Evacuating the sealed container (fruit juice)

(iii) Use of oils from top of the food (pickles)

4. Exposing at high temperature: Fruits can be exposed by high temperature such as:

(i) Canning: Food is exposing to a high temperature (> 100^0C) which prevents spoilage and inactivate the enzyme present in the food.

(ii) Irradiation: In case of irradiation, the food is exposed to the radiations to kill the micro organism by ionizing and non-ionizing radiation like α, β and gamma rays. Her, food is exposed to electromagnetic or ionizing radiation or various frequencies ranging from low frequency electromagnetic to high frequency *i.e.,* gamma rays which destroy the micro-organism present in the food.

B. Prevention/delay the shelf decomposition

(i) By destruction or inactivate the enzyme - Blanching.

(ii) Prevention / delay the non-enzymatic chemical reactions - Antioxident

Blanching

1. It is primary treatments which have to soften the tissues to facilitate packaging.
2. To preserve the original colour and flavour
3. To destroy the certain enzyme which are undersirable
4. Elimination of the air
5. Mostly for vegetables
6. Remove micro-organisms
7. Remove astringent taste and toxins

Anti-oxidant

Anti-oxidant are substances which are used to protect the food gamma deterioration caused by exposure to the air.

1. BHA - Butylactic Hydroxy Anisole Vegetable oils

   BHT - Butylactic Hydroxy Toluene
2. Gellales: Animal fat, Vegetalbe oil
3. Tocopherols: Animal fat
4. Ascorbic acid: Fruit juices, Citrus oil, Wine, Bears *etc.*
5. Lactic acid: Processed fruits and vegetables, Canned fruits,
6. Phosphoric acid: Vegetable oils, Animal fat and Cola drinks

Methods of Preservation of Fruit and Vegetable

There are two main basic methods:

A. Bacteriostatic methods

B. Bactericidal methods

A. Bacteriostatic methods

1. Drying of foods
2. Use of chemical preservatives
3. Use of food additive
4. Use of low temperature

B. Bactericidal methods

1. Pasteurization
2. Cooking
3. Canning
4. Irradiation

A. Bacteriostatic Methods

In this method, the environmental conditions are change to prevent the growth of micro organisms, such conditions are called bacteriostic. These are:

1. Drying of Foods

Drying is just removal of moisture from the food to a certain level at which micro organisms cannot grow is called drying, it can be done by two methods:

(i) Application of heat

(a) Sun drying (b) Mechanical drying
(c) Vacuum drying (d) Freeze drying

(ii) Binding the moisture in the food

(a) Use of Sugar (b) Use of Salt

(i) Application of heat

a) **Sun Drying:** Sun drying is the method in which food is directly exposed to sunlight. It is generally done in the places where plenty sunshine is available for long period e.g., Rajasthan. The dried product in this method is inferior in quality.

b) **Mechanical drying:** This is a method of drying where application of heat is applied by a mechanical dryer under the controlled conditions of temperature, humidity and air flow.

c) **Vacuum drying:** The temperature of the food and the rate of water removal are controlled by regulating the degree of vacuum and intensity of heat input.

d) **Freeze drying:** In this method, the food is dried by sublimation process, *i.e.,* just converting the food into ice without passing through the liquid form of water by means of vacuum plus heat applied in the drying chamber. In this method, product first frozen then water is removed by vacuum and application of heat which occurs simultaneously in same chamber.

(ii) Binding the moisture

a) **Use of Sugar:** The use of high concentration of sugar bindup the moisture and make the food have a certain level of moisture at which micro organisms are not able to grow.

b) **Use of Salt:** The concentration of salt causes the high osmotic pressure and tie up the moisture which inhibit the growth of micro organisms. It dehydrates the food by drying out and tie up moisture as it dehydrate the micro organisms cells. Salt reduces the solubility of O_2 in the food by reducing the moisture. It interferes with the action of proteolytic enzyme. The effeteness of NaC1 is varied with the concentration of salt and temperature.

2. Use of Chemical Preservative

Chemical preservatives are substances which are added to food just to retard, inhibit or arrest the activity of micro organisms such as fermentation, pacification and decomposition of the food. Chemical preservatives are of two types:

Class-1 preservatives: Common salt, sugar, dextrose, spices, vinegar, ascorbic acid

Class-2 preservatives: Benzoic acid and its salt, SO_2 and the salts of sulphuric acid, nitrates, sorbic acid and its salts, propeonic acid and its salts, lactic acid and its salts.

Among the class-2 preservatives, only two chemical preservatives are used in fruits and vegetables Preservation:

(i) KMS

(1) It releases the SO_2 and it is unstable.

(2) It is used for the fruit which have non water solvent pigment (colourless).

(3) It cannot be used in naturally coloured juices such as phalsa, jamun because they have the Anthocynin pigment.

(4) It cannot be used in the product which are packed in container because it acts on the tin containers and oil, Hydrogen Sulphide (H_2S) which has an unpleasant smell and also form a black compound with the base plate of containers.

(5) Best to control moulds than bacteria.

(6) 350 ppm KMS is mostly used in fruit juice products.

(ii) Sodium Benzoate

(1) It is salt of benzoic acid and soluble in water.

(2) It delays the fermentation in the juices.

(3) It is commonly used in the product which are having natural colour such as anthocynin pigment.

(4) It is more effective against the yeast.

(5) 750 ppm Sodium benzoate is mostly used in fruit juices, squashes and cordials.

3. Use of Food Additive

Food additives are substances or mixture of substances other than basic foodstuffs, which are present in the foods as reagent of any aspects of production, processing, storage, packaging *etc.* Food additives are:

(i) Sugar,

(ii) Salt,

(iii) Acids,

(iv) Spices

In case of sugar and salts, they exerts osmotic pressure by water is diffuses from the product through a semi-permeable membrane until the concentration reached equilibrium. They kills the micro organisms or do not allow them to multiplication.

(i) **Sugar:** The concentration of 68-70% is used for preparation of jam, jelly, marmalades *etc.* sugar act as a preservative by osmosis and not as a true poison for micro organisms. It absorbs most of the available water, so little water available for the growth of micro organisms.

(ii) **Salt:** the concentration of salt 15-20% is used for the preparation such as pickles. Salt inhibits enzymatic browning and discolouration and also acts as an anti-oxidant.

It exerts its preservative action by:

1. Causing high osmotic pressure resulting in the plasmolysis of microbial cells.
2. Dehydrating food and micro organisms by tying up the moisture.
3. Ionizing to yield the chloride ion which is harmful to micro organisms, and
4. Reducing the solubility of oxygen in water, sensitizing the cells against CO_2.

(iii) **Acids:**

1. Many processed foods and beverages need the addition of acids to impart their characteristic flavour and taste in the final product because an acid provides desired flavour and taste.
2. They adjust the sugar and acid ratio in the food.
3. Proper balance flavour of the food.
4. They are also playing the role for controlling the pectin-gel formation.

Main acids are the following :

1. Acetic acid (Vinegar)
2. Citric acid (Lime juice)
3. Lactic acid (Lactose)

1. Acetic acid: It is commonly used for pickles, chutney, sauce and ketchup, just to inhibit the growth of micro organisms.

2. Citric acid: It is used for preparation of jam, jelly, squash, nectar *etc.* just to increase the acidity.

3. Lactic acid: It is used for the formation of curd from milk, raw flavour, specific to pickles

(iv) Spices:

(1) Spices are plant products which are used in flavouring the foods and beverages to enhance the food flavour, colour and palatability.

(2) They act as antibacterial and antifungal activity.

(3) They impart as colour agent.

4. Use of Low Temperature

Low temperature retards the microbial growth and enzyme reaction because it retards the chemical reactions. This is not a permanent method because some micro organisms can also grow at low temperature.

1.	Cellar storage	:	(Above 15^0C)
2.	Refrigerated storage	:	(0 to 5^0C)
3.	Freezing storage	:	(-18 to -40^0C)

1. Cellar Storage: These are the underground room where surplus food can be stored for some time; only root crops such as potato, onion can be stored for a limited period.

2. Refrigeration: Fruits and vegetables can be stored for 2-7 days. Semi-perishable crops, such as potatoes, apples *etc.* can be stored, in the commercial cold storage with proper ventilation, automatic controlled temperature for one year.

3. Freezing: It tie up the moisture and increase the concentration of dissolved substances in the food. But, sometimes enzymes are active even below the 0^0C.

In this case before freezing, 'Blanching' is necessary for vegetable freezing.

B. Bactericidal Method

In this method, food material is exposed to higher temperature and high temperature helps to killing of the micro organisms due to coagulation of protein. It helps in inactivation of enzyme. Here moist heat is more effective than dry heat. At low pH high temperature is required than the high pH. High temperature can be employed by following methods:

(i)	Pasteurization	:	Below 100^{0}C
(ii)	Boiling/ Cooking	:	at 100^{0}C
(iii)	Canning	:	Above 100^{0}C

(i) Pasteurization

There are three methods of pasteurization

a) Bottle or holding pasteurization

This method is commonly used for the preservation of fruit juices at home. The extracted juice is strained and filled in bottles, leaving sufficient head space for the expansion of the juice during heating. The bottles are then sealed air-tight and pasteurized.

b) Overflow method

Juice is heated to temperature of about 2.5^{0}C higher than the pasteurization temperature and then filled in hot sterilized bottles up to the brim, during filling and sealing the temperature of juice should not fall below the pasteurization temperature.

c) Flash pasteurization

The juice is heated rapidly to a temperature of about 5.5^{0}C higher than the pasteurization temperature and kept at this temperature for about a minute. This method commonly used for canning of natural orange juice, grape and apple juices. It is a mild heat treatment; by pasteurization milk is pasteurized by HTST at 72^{0}C for 15 sec. Fruit juices are pasteurized at such temperature and for such periods as would render them sterile, without impairing their flavour. Usually, the juices are pasteurized at about 85^{0}C for 25-30 min., according to the nature of the juice and the size of container. Acid fruit juices require lower temperature and less time for pasteurization than the less acid ones.

Juices can be pasteurized in two ways :

(1) By heating the juice at a low temperature for a High time (LTHT) and

(2) By heating the juice at high temperature for a short time (HTST).

(ii) Boiling/Cooking

The primary objective of cooking is to produce a palatable food. Cooking results in:

1. Destruction or reduction of micro organisms and inactivation of undesirable enzymes.
2. Destruction of potential hazard in the foods which are present naturally through micro-organism.
3. Improvement of colour, flavour and texture of the food.
4. It improves the digestibility of food component.
5. Putting the temperature about 100°C. by this method, food can be preserved for 10-24 hours at low temperature.

(iii) Canning

Canning is done at or above 100^0C. In case of fruits which are acidic, they are canned at 100^0C, while in case of vegetable those are nonacidic, they are canned at above 100^0C. Here, high temperature can be obtained by using steam pressure; time is varying according to the type of foods. Due to anaerobic condition any survivable organism will not grow.

On the basis of Acid, foods are divided into four different groups:

1. Low Acid Foods (pH 5.3 and above): Peas, Corn, Lime beans, Meat, fish, Poultry and Milk.
2. Medium Acid Foods (pH 5.3-4.5): Spinach, Asparagus, Beets and Pumpkin.
3. Acid Foods (pH 4.5-3.7): Tomatoes, Pears and Pineapple, Sauce.
4. High Acid Foods (pH below 3.7): Berries and Sauer kraut, Pickle.

Other Methods of Preservation

Preservation by filtration

In this method juices are clarified by settling or by using ordinary filters, and then passed through special filters which are capable of retaining yeasts and bacteria. Various types of germ-proof filters are used for this purpose.

Preservation by carbonation

Carbonation is the process of dissolving sufficient CO_2 in water or beverage so that the product when served gives off the gas as fine bubbles and has a characteristic taste, carbonation life of beverage. Fruit juice beverages are generally bottled with CO_2 content varying from 1-8 g/lit though carbon should be avoided as it destroys the flavour of the juice. The keeping quality of carbonated fruit beverages is enhanced by adding about 0.005% sodium benzoate. The level of carbonation is required according to the type of fruit juice and type of flavour.

Preservation by fermentation

This is one of the oldest methods of preservation. By this method the foods are preserved by the alcohol or organic acid formed by microbial action. The keeping quality of alcoholic beverages vinegars and fermented pickles depends on the presence of alcohol acetic acid and lactic acid respectively. Wines, beers, vinegar, fermented drinks, fermented pickles are prepared by these processes. Fermentation is carried out by using acetic acid, lactic acid and alcohol *etc.*

Preservation by antibiotics

Certain metabolic products of micro-organisms which are found to have a germicidal effect those are:

Nissin: - is an antibiotic produced by *Streptococcus lactis* an organism found in milk, curd, cheese and other fermented milk products. It is non toxic and it is widely used in food industry especially for preservation of acid foods.

It is commonly used in canning of mushroom tomatoes and milk products. Nisin suppress the growth of spoilage organisms.

Subtilin: - an antibiotic obtained from *Bacillus subtilis* is used in preservation of asparagus, corn and peas.

Pimaricin: - an antifungal antibiotic used for treating fruits and fruit juices.

Preservation by irradiation

It is a process of preservation of food by exposing them to ionizing energy radiation which kills most of the spoilage causing organism and also inactivates the enzymes responsible for browning *etc.* This method prevents the sprouting in storage condition (onion, potato *etc.*).

The irradiation of food can be considered to be a method of "Cold sterilization".

Points to Remember

- The oxicity of SO_2 is increased at hight temperature.
- SO_2 content in pure KMS is 57.7%.
- Commercial effect of KMS is due to H_2SO_4.
- Germicidal effect of KMS is due to H_2SO_3.
- Dehydration and drying is permanent method of preservation.
- Sterilization by Ultra-high temperature (UHT) is done at temperature of 149°C for 2 second.
- In carbonated fruit juice beverages CO_2 content is 1-8g/liter.
- Cold process is also known as Ultrafiltration.

❑❑❑

14

Food Additives

Food additive is any chemical substance that is added to food during preparation or storage. A food additive either becomes a part of the food or affects its characteristics for achieving a particular effect. For instance, substances that are used to enhance the appearance, texture, or to keep qualities of a food are considered food additives.

Function of Food Additive

1. Give the food a smooth and consistent texture

- Emulsifiers prevent products from separating.
- Stabilizers and thickeners provide an even texture.
- Anticaking agents allow substances to flow freely.

2. Improve or preserve the nutrient value

- Many foods and drinks are fortified and enriched to provide vitamins, minerals, and other nutrients to many foods, such as flour, cereal, margarine, and milk.
- This helps make up for vitamins or minerals that may be low or lacking in a person's diet.
- All products that contain added nutrients must be labeled.

3. Maintain the wholesomeness of foods

- Bacteria and other germs can lead to foodborne illnesses. Preservatives reduce the spoilage that air, fungi, bacteria, or yeast can cause.
- Certain preservatives help preserve the flavor in baked goods by preventing the fats and oils from going bad.
- They also keep fresh fruits from turning brown when exposed to the air.

4. Control the acid-base balance of foods and provide leavening

- Certain additives help change the acid-base balance of foods to get a certain flavor or color.
- Leavening agents that release acids when they are heated react with baking soda to help biscuits, cakes, and other baked goods rise.

5. Provide color and enhance flavor

- Certain colors improve the appearance of foods.
- Many spices, as well as natural and man-made flavors, bring out the taste of food.

Types of Food Additive

Acids and alkalies

These food additives are added to make flavors "sharper", and also act as preservatives and antioxidants. Common food acids include vinegar, citric acid, tartaric acid, malic acid, fumaric acid, ammonium carbonate, calcium citrae, calcium carbonate and lactic acid.

Acidity regulators

Acidity regulators are substances added to food to function as preservatives or flavor agents. Acidity regulators work by altering or stabilizing the pH of a food. They differ from acidifiers in that acidifiers work to lower pH, here acidity regulators work to stabilize pH. Common food acidity regulators include fumaric acid, aluminum potassium sulfate, calcium gluconate, potassium gluconate, potassium acetate, malic acid and acetic acid.

Acidifiers

Acidifiers are substances added to food as preservatives and flavor agents. They work to bring the pH level down towards acidity. Acidifiers like lactic acid work as anti-microbial agents. Others like citric acid can function to protect fats in food from oxidizing. Common food acidifiers are lactic acid, ascorbic acid, citric acid and phosphoric acid. Mostly acidifier used in canned fruit, soda, soups, sauces, beer.

Acidulate

Acidulate is food additives that alter the pH of food in order to yield acidity and sourness to flavor. Common food acidulates are acetic acid, citric acid, phosphoric acid, tamarind, fumaric acid, sodium benzoate and metal bisulfates. Commonly acidulate are used in pureed foods, candy, condiments.

Anti-browning agents

Anti-browning agents are additives that function to prevent browning in foods such as fruit or meat. Common food Anti-browning agents are sodium sulfite, citric acid, cysteine, potassium sulfite, sulfer dioxide, sodium metabisulfite and oxyresveratrol. Commonly anti-browning agents used in meats, fruits, vegetables.

Anti-caking agent

Anti-caking agents are additives absorb moisture to prevent powdered or granulate foods from lumping together or caking. Common anti-caking agents are aluminum silicate, sodium silicate, calcium silicate, magnesium silicate, calcium carbonate, magnesium carbonate, powdered cellulose and ferric ammonium citrate. Commonly anti-caking agents used in salt, flour, sugar, coffee.

Antioxidants

Antioxidants act as oxygen scavengers as the presence of oxygen in the food helps the bacteria to grow that ultimately harm the food. In the absence of anti oxidant food additive oxidation of unsaturated fats takes place rendering to foul smell and discoloration of food. Different kinds of antioxidants foods act in a different ways but the end result is to delay or minimize the process of oxidation in food. Some antioxidants foods additives combine with oxygen to prevent oxidation and other prevent the oxygen from reacting with the food leading to its spoilage.

eg.- ascorbic acid, citric acid, tocopherols, butylated hydroxyanisole (BHA), butylated hydroxytolune (BHT), tertiary-butylhydroquinone (TBHQ) and propyl gallate (PG).

Bleaching agents

Bleaching agents are bleach or whiten flour and bread. Eg., benzoyl peroxide, calcium peroxide, hydrogen peroxide, sodium metabisulfite, etc.

Bulking agents

Bulking agents are non caloric additives used to impart volume and mass to a food. Soluble fibers such as guar gum and psyllium hust are one of the most common forms of bulking agents. Common bulking agents are guar gum, psyllium husk, carnuba wax, mannitol, maltitol, glycerin, methylcellulose, polydextrose, beta glucan and pectin. Commonly bulking agents used meal replacements, low calorie foods, cereals, pastries, breads, most processed food.

Colouring agents

These are added to a large number of food item to make them attractive and appetizing. They may be of natural origin, e.g., extract of annatto, caramel, carotene and saffron or synthetic, e.g., carbon black, etc.

Clarifying agents

These are used to remove haziness or sediment produced by oxidative deterioration in fruit juice, wines, beers, etc. E.g., Bentonite, Gelatin, Synthetic resins (Polyamides and polyvinylpyrrolidone).

Enzymes

Enzymes are capable of catalysing a specific biochemical reaction and added to make improvements in the quality of different foods (at any stage of the manufacturing, processing, preparation, treatment, packaging, transport or storage of foods). E.g., amyloglucosidase, carbohydrase, α-galactosidase, α-amylase, etc.

Firming agents

Firming agents are add to keep the tissues of fruits and vegetables crisp (firm), e.g., Calcium chloride and aluminium sulphate.

Humectants

Humectants prevent foods from drying out e.g., Propylene glycol, glycerol sorbitol and mannitol.

Flavouring agents/Flavour enhancers

Flavors are additives that give food a particular taste or smell, and may be derived from natural ingredients or created artificially and flavor enhancers enhance a food's existing flavors. They may be extracted from natural sources (through distillation, solvent extraction, maceration, among other methods) or created artificially.

Nutrient supplements

Nutrient supplements provides extra nutrients in food for consumer e.g., vitamins and minerals.

Preservative

Any substance which is capable of inhibiting, retarding or arresting the growth of microorganism is known as preservative. It may be a chemical or a natural substance. These are two types that are under:

Class I preservatives including vinegar, salt, sugar, honey, vegetable oil.

Class II preservatives are chemicals like sodium benzoates, sorbates and potassium metabisulfites.

Sequestrants/chelating agents/metal scavenger

These are combine with metals such as iron and copper and remove them from solution. E.g., Ethylenediamine tetraacetic acid (EDTA), polyphosphate or citric acid, calcium chloride, calcium phosphate.

Stabilizer and thickeners

Thickeners are substances which, when added to the mixture, increase its viscosity without substantially modifying its other properties, they help to stabilize emulsions e.g., Gum Arabic, agar-agar, alginic acid, starch and its derivatives, gelatin, pectin, amylase, carboxyymethylcellulose (CMC) carrageenan, hydrolyzed vegetable proteins, etc.

Surface active agents/emulsifiers

Emulsifiers allow water and oils to remain mixed together in an emulsion, as in mayonnaise, ice cream, and homogenized milk e.g., Propylene glycol monostearate, monosodium phosphate.

Sweeteners

These are sweetening agents other than sugar. They are widely used in low calorie products to aid slimming and, at the same time, provide a sweet taste.

Points to Remember

- Tastes are four types: - Salty, Sweet, Sour and Bitter.
- Sodium Chloride (NaCl) is the only salt that has pure salty taste.
- Iodides and Bromides are bitter in taste.
- Lead and Beryllium salt are sweet in taste.
- Sodium Chloride not only use in flavour in food but also essential nutrient.
- Sugars are used for sweetness of food.
- Fructose is sweetest sugar followed by sucrose and glucose.
- Natural sweet compounds are generally polyhydroxy compound with a straight chain structure e.g. Sugar, mannitol, sorbitol etc.
- Sucrose found in sugar cane and root of sugar beet.
- Maltose found in barley and during the germination of certain grain.
- Lactose is the naturally occurring sugar found in milk.
- Fructose mostly found in fruits.
- Lactose is found in milk.
- In food sourness due to the presence of organic acid (citric acid, tartaric acid, malic acid and oxalic acid).
- Hydrogen ions are mainly responsible for sour taste.
- Bitterness in food may be due to alkaloids, glycosides and other organic compound as well as inorganic salts.
- Phosphoric acid is used in soft drinks.

15

Canning

Canning is a process of sealing food stuffs hermatically in containers and sterilizing them by heat for long storage is known as canning.

Principle of Canning

Destruction of spoilage organism within the sealed container by means of heat.

Selection of Fruit and Vegetable

For canning fruit should be ripe but firm and uniformity mature and vegetables except tomato should be tender.

Grading

Selected fruit and vegetable are graded according to size and colour to obtain uniform quality. This is done by hand or machines such as screw grader and roller grader.

Washing

It is important to remove pesticide spray residue and dust from fruits and vegetables.

Peeling

Peeling is to remove the outer layer of fruits and vegetable. It is done in various ways that are:-

a) **Hand peeling:** It is done in case of fruits of irregular shape. *e.g.* Mango and Papaya where mechanical peeling is not possible.

b) **Steam peeling:** Peaches are steam peeled in different ways, potatoes and tomatoes are peeled by steam or boiling of water.

c) **Mechanical peeling:** This is done in case of apples, peaches, pineapples and cherries and also for root vegetable like carrots, turnips and potatoes.

d) **Lye peeling:** fruits like peaches, apricots, sweet oranges, mandarin oranges and vegetable like carrots and sweet potatoes are peeled by dipping them in 1-2% boiling caustic soda solution (lye) for 30 second – 2 minute depending on their nature and maturity.

e) **Flame peeling:** It is used only for garlic and onion which have a papery outer covering. This is just burnt off. Vegetable like peas are shelled, carrots are scraped, and beans are snipped or trimmed.

Cutting

Pieces of the size required for canning are cut, seed, stone and core are removed.

Blanching

Heat treatment of vegetables in boiling water or steam for 2-5 minute prior to canning for inactivation of enzymes, removal of air, setting of natural colour, improvement of texture, arresting change in flavour, softening of the products, etc.

Syruping

A solution of sugar in water is called as syrup which 20° to 30° brix is used for canning of fruits.

Cooling

After blanching the vegetable are dipped in cold water for better handling and keep them in good condition.

Filling

Before filling cans are washed with hot water and sterilized. But in developing countries these are subjected to a jet of steam to remove dust and foreign material. Automatic large can filling machines are used in advanced countries. After filling covering with syrup or brine is done and this process is called syruping or brining.

Exhausting

The process of removal of air from cans is known as exhausting. After filling and lidding / clinching, exhausting is essential.

Advantages of Exhausting

a) Corrosion of tinplate and pin-holing during storage is avoided.

b) Minimizes discolouration by preventing oxidation.

c) Retention of vitamin-C.

d) Reduces chemical reaction between the container and the contents.

e) Prevents development of excessive pressure.

Sealing

Immediately after exhausting the cans are sealed airtight by means of can sealer. In case of glass jars a rubber ring should be placed between the month of the jar and the lid, so that it can be sealed airtight, during sealing the temperature should not fall below 74^{0}C.

Flow chart for canning process

Selection of fruit and vegetable

↓

Grading

↓

Washing

↓

Peeling

↓

Cutting

↓

Blanching

↓

Cooling

↓

Filling and syruping/brining

↓

Exhausting

↓

Sealing

↓

Processing

↓

Cooling

↓

Storage

Processing

Heating of foods for preserving is known as processing. In the canning technology processing means heating or cooling of canned foods to inactivate microorganisms. Many bacterial spores can be filled by either high or low temperature. Processing time and temperature should be adequate to eliminate all bacterial growth. Almost all fruits and acidic vegetables like tomato and rhubarb are processed at a temperature of 100°C and vegetables are processed at a temperature of 115 to 121°C.

Heat processing requirements for different products based on different acidity and pH

Acidity class of food	pH	Food item	Processing requirements
Low acid	5.0 – 7.0	Peas, carrots, beets, potatoes, asparagus, tomato soup	High temperature processing 116-121°C (240-250°F)
Medium acid	3.7 - 4.5	Tomatoes, pears, apricots, peaches	-do-
Acid	3.0 - 3.7	Sauerkraut, pineapple, apple	Boiling water processing 100°C (212°F)
High acid	3.0 – 2.3	Pickles, lime juice	-do-

Cooling

After processing, the cans are cooled rapidly to about 39^{0}C to stop the cooking process and to prevent stack-burning. Cooking is done by the following methods:-

a) Dipping or immersing the hot cans in tanks containing cold water (Generally this method is used)

b) Letting cold water in to pressure cooker specially in case of vegetable

c) Spraying cans with jets of cold water

d) Exposing the cans to air.

Storage

After labeling the cans, they should be packed in strong wooden cases or corrugated cardboard cartons and stored in a cool and dry place. Storage of cans at high temperature should be avoided as it shortens the shelf-life of the product.

Type of Cans

Two type of cans are use in canning of fruit and vegetables that are:

1. **Tin containers:** Tin cans are made of thin steel plate of low carbon content, lightly coated on both side with tin metal. It is necessary, therefore, coating of inside the cans with some material to prevent corrosion of the plate and discoloration of the product.

   Two types of lacquers are used:

   I. **Acid-resistant (R-enamel/A.R.Cans):** cans are used for packing fruits (acid food material) like pineapple, peach, apricot, grapefruit, strawberry, grape etc.

   II. **Sulphur-resistant (C-enamel/S.R.Cans):** cans are used for packing vegetables (non-acid food material) like pea, corn, limba beans etc.

2. **Glass containers**

Points to Remember

- Nicholas Appart (1804) is known as father of canning.
- Canning is also known as Appertizing.
- Fruits like berries, plum and cherries are graded whole, while peach, pear, apricot, mango, pineapple etc. are generally graded after cutting into pieces or slices.
- Shelling is done in pea for canning.
- Snipping " trimming are done in beans (removal of the fibrous string) for canning.
- Scrapping is done carrot (epicarp is scrapped) for canning.
- Exhausting is practiced at 82-87°C temperature so that the temperature in the centre of the can may reach to79°C.
- During sealing the temperature should not fall below 74°C.
- Heating or cooling food to inactive bacteria is termed as precooling.
- Cans should have 0.3-0.5 cm head space, it should not be more than 1.6cm.
- Drained weight of the fruit should not be less than 50% and 55% in vegetables.
- No artificial colour and preservative shall be added except pea.
- Canned fruits like peach, grape fruit, pineapple, bean, spinach, pea, celerary can be stored for more than 2 years.
- Apricot should not be peeled for canning.
- Fruits in which salt added before canning are apple and guava.
- At high storage temperature hydrogen formation will be more, hence canned products should be stored under cool and dry condition.
- Fruit Preservation and Canning Institute was established at Lucknow in 1949 by the Government of Uttar Pradesh.
- Hydrogen swell is due to the hydrogen gas produced by the action of food acids on the metal of the can.
- Citric acid (0.5%) checks the formation of hydrogen swell to a great extent.

- Panelling is generally seen in large sized can that the body is pushed inward due to the high vacuum inside.
- Bursting caused by the excess pressure of gases produced by decomposition of the food by microorganisms or by hydrogen gas formed by the chemical action of food acids on the tin plate.
- Enzymatic browning develop in cans due to the enzyme polyphenol oxidase present in fruits and vegetables, it can be avoided by placing the peeled and cut pieces in 2% salt solution.
- Tomato are harvested at red ripe stage for canning.
- Blanching is important operation for vegetable drying and dehydration but during blanching 10-50% vitamin-C loss.
- Thermal processing of ethylene production, mango can be classified into SR cans.
- Metallic contaminations are mainly due to iron and copper salts, it can be avoided by using glass containers, coating the interiors of can with lacquer and also eliminating the use of iron and copper vessels.
- Lacquering: - coating of inside the cans with some material to prevent corrosion of the plate and discoloration of the product.
- Plains cans are used for fruit which are insoluble color.
- Pink colorations is a problem in canned cabbage and peach is due to low cooling.
- In canning flat sour spoilage is mostly in non acid vegetables caused by *Bacillus coagulans* and *Bacillus sterothermophilus*, which produce acid without formation of gas.
- Copper is used to fix green colour in beans.
- Stack burning is due to poor cooling of the cans.
- Thermophilic acid spoilage, the cans swell due to production of carbon dioxide and hydrogen by *Clostridium thermosacccharolyticum*. This type of spoilage mostly occure in low and medium acid foods.
- Sulphide spoilage is caused by *Clostridium nigrificanns* in low acid foods.
- *Clostridium butyricum* and *C. pasteurianum* cause a butyric acid type of fermentation in foods with swelling of the container due to the formation of carbon dioxide and hydrogen.

- Breathing is due to presence of very tiny leak in the cans.
- Lacqured cans are used for beans, asparagus, pea and pumpkin.
- Kew variety of pineapple suitable for canning.
- Toughening effect on canned bean is due to Ca.
- Processing temperature for canning of fruits is 100°C and for vegetables is 115-121°C.
- 120 tendrometer reading should be foe canning of pea.
- In canned fruits, drained weight should not be less than 50%.
- According to FPO, permissible limit of Tin content in process food is 5 ppm.
- In, Anola and carrot peeling is done by Lye peeling.
- Lay peeling is done at 93°C temperature.

16

Drying

Removal of water from the food by non-conventional energy source like sunlight, wind, and a smoky fire. Drying (dehydrating) food is one of the oldest and easiest methods of food preservation.

Dehydration

Dehydration is the process of removing water or moisture from a food product under controlled condition like temperature, relative humidity, air flow, etc.

Principles of Drying and Dehydration

Drying removes the moisture from the food so that bacteria, yeasts and moulds cannot grow and spoil the food. It also slows down the action of enzymes, but does not inactivate them. Dried food items can be kept almost indefinitely, as long as they are not rehydrated.

Objective of Drying and Dehydration

1. To reduce the bulk and weight ($1/4^{th}$ to $1/9^{th}$ of the fresh weight).
2. To reduce the water activity (a_w).

Factor Affecting the Drying Rate

a) Composition of raw material

b) Size, shape and arrangement of stacking of produce

c) Temperature, humidity and velocity of air
d) Pressure (barometric or under vacuum)
e) Heat transfer to surface (conductive, convective / irradiative)
f) Relative humidity of the air

Effect of Food Properties on Dehydration

Drying techniques / methods

A) Fruit and vegetable natural drying

1) **Sun or solar drying:** Surplus production and specially grown crops may be preserved by natural drying for use until the next crop can be grown and harvested. These can be cheaply distributed t areas where there are permanent shortages of fruit and vegetables.
2) **Shade drying:** It is done for products which can lose their colour and or turn brown if put in direct sunlight. Eg: Herbs, Green and red sweet peppers, chillies, green beans and okra.
3) **Osmotic dehydration:** Here the prepared fresh material is soaked in a heavy or thick liquid sugar solution or strong salt solution and then the material is solar dried.

B) Common driers used for drying / dehydration

Drier types	Food types
a) Air convection driers	
1) Kiln	Pieces
2) Cabinet, tray or pan	Pieces, purees, liquid
3) Tunnel	Pieces
4) Continuous conveyor belt	Pieces, purees, liquids
5) Belt through	Pieces
b) Drier type	
1) Air lift	Small pieces granules
2) Fluidized bed	Small pieces, granules
3) Spray	Liquid purees
c) Drum or roller driers	
1) Atmospheric	Purees, liquids
2) Vacuum	Purees, liquids

Contd.

d) Vacuum driers	
1) Vacuum shelf	Pieces, purees, liquid
2) Vacuum belt	Purees, liquids
3) Freeze driers	Pieces, liquids

Change during Drying

Pre-Dehydration Treatments for Drying

1. Blanching

It is a pre-cooling treatment in which vegetables are usually heated in boiling water for 2-5 minute to inactivate the natural enzymes before processing. Blanching is essential for all vegetables except onions, peppers, okra, herbs, and some new types of corn that get sweeter as they mature.

Blanching helps save some of the vitamin content, sets the color, hastens drying by relaxing the tissues, and helps food reconstitute.

2. Sulphuring

Exposure of the prepared fruit pieces to burnt sulfur fumes before drying which helps to maintain and attractive light colour, conserves vitamins-C and perhaps A, make drying process eaiser by softening cell wall, displaces air form the tissue, destroys enzymes that cause darkening of cut surface and kills many of the micro-organisms present.

Post-Dehydration Treatments for Drying

1. Sweating

Sweating is a process of equalization of moisture to a desired level of the dried fruit and nuts (almond, walnut) before packaging.

2. Heat treatment

Dried fruits and vegetables are generally attacked by insect even when they are properly dried and stored. Insect not only consume the material but also leave the debris which spoiled the appearance of product. To avoid insect infestion, great care in the construction of practices on godown. In case of heat treatment, dried fruit such as rasins, fig, peaches are dipped in water for dilute solution of NaCl or Na_2HCO_3 and then redried at 54-65^{0}C to destroy all insects including their formative stages. Dried vegetable may be heated such directly without any preliminary treatments. The heat sterilized materials should be packed in clean and sterilized container wich are insect proof.

3. Fumigation

Treatment of materials in a close chamber by a fumigant to kill the dry storage pest. Mostly Methyl Bromide is use as a fumigant in dry products storage.

Intermediate-Moisture Foods (IMF)

Intermediate-moisture food or semimoist foods, contain moderate levels of moisture, of the order of 20-50% by weight, which is less than is normally present in natural fruits and vegetables, but more than is left in conventionally dehydrated products. In addition, Intermediate-moisture food contains sufficient dissolved solute to decrease water activity below that required to support microbial growth.

Points to Remember

- Vegetable are generally dried at 60-66^{0}C and fruits at 66-71^{0}C.
- Relation between moisture and temperature is called psychrometric relation.
- Final moisture content in dried vegetable is 6-8% and for fruit 10-20%.
- Sulphuring done in mostly for fruits.
- Calcium use as a desiccant in packaging.
- Pea should have the tendrometer reading 100-105 for dehydration.
- Dried banana known as a banana fig.
- In soft date invert sugar is predominate.
- Sugar content in grape raisin should be 20-24^{0}B.
- Alkaline dip (0.5%) has been given to cranberry for their better drying.
- Water, glycerol, propyl glycol use as plasticizer.
- Methyl cellulose use as a foaming agent for apple, peach, cherry drying.
- Propylene GMS use as a foaming agent for lemonade.

- Symarn fig is suitable for drying.
- Kiln drier will not reduce moisture to below about 10%. It is use for apple slices and occasionally for potato.
- Temperature in Belt Trough Drier: 135^0C.
- Temperature in Cabinet, Tray and Pan Drier: 95^0C and air velocities are: 2.5-5 m/sec.
- Freeze-Drying two to five times greater per weight of water removed then other common drying methods.
- Most common preservation method for grapes is drying.
- Freeze-Drying is use for high value foods such as coffee, fruit juice, strawberries and mushroom slices.
- In Freeze-Drying the material dries directly by sublimation of ice without passing through the intermediate liquid stage.
- The Freeze-Drying products are highly hygroscopic and it reconstitutes readily.
- Moisture content of Freeze-Drying products is 2-3%.
- In dried product browning occur due to Maillard reaction or enzyme catalyzed reaction it can be avoid by store at low temperature and enzyme inactivation by blanching or steaming before drying.
- Blanching is done in vegetable for drying.
- Onion is not blanched before drying.
- Carometization of reducing sugar is the main reason for browning in fruits when dehydrated at high temperature.
- Tenderometer reading should be 85-95 for drying of pea.
- The point at which the dried products just become lumpy is known as critical point.
- White colour onion are used for dehydration.
- The point at which the dried products just become lumpy is known as critical point.
- The point which have 5% low relative humidity than the critical point known as danger point.

Flow sheet for drying / dehydration of fruits and vegetables

Fruit and vegetable
(Mature and free from insects and disease)

↓

Washing

↓

Peeling

↓

Preparation

↓

Blanching

↓

Spreading on flat-wooden trays

↓

Sulphuring
(usually @ 1.8-3.6 kg/ton of fruits)

↓

Drying/ Dehydration
(with occasional stirring)

↓

Sweating

↓

Packing
(Air tight container or polythene bags)

↓

Storage
(At ambient temperature in dry place)

Blanching and Drying Times for Selected Fruits

Fruit	Blanching		Drying time(hrs)
	Method	Time (mins)	
Apple	steam	3–5	6–12
	syrup	10	
Apricots	steam	3–4	24–36+
	syrup	10	
Bananas	steam	3–4	8–10
	syrup	10	
Cherries	syrup	10	24–36
Figs	not necessary	6–12	
Grapes: seedless	not necessary	12–20	
Nectarines	steam	8	36–48
	syrup	10	
Peaches	steam	8	36–48
	syrup	10	
Pears	steam	6	24–36+
	syrup	10	
Pineapples	not necessary	24–36	
Plums	not necessary	24–36	

Drying Ratio of Vegetables

Vegetable	Ratio (Fresh:dried)
Bean and Potato	7:1
Onion	10:1
Okra	12:1
Bitter gourd	16:1
Cabbage and carrot	18:1
Pumpkin and spinach	22:1
Tomato	27:1
Turnip	28:1
Brinjal	33:1
Cauliflower	35:1

Blanching and Drying Times for Selected Vegetables

Vegetable	Blanching		Drying time(hrs)
	Method	Time (mins)	
Beets	cook before drying	3½–5	
Carrots	steam	3–3½	3½–5
	water	3½	
Corn	not necessary	6–8	
Garlic	not necessary	6–8	
Horseradish	not necessary	4–10	
Mushrooms	not necessary	8–10	
Okra	not necessary	8–10	
Onions	not necessary	3–6	
Parsley	not necessary	1–2	
Peas	steam	3	8–10
	water	2	
Peppers	not necessary	2½–5	
Potatoes	steam	6–8	8–12
	water	5–6	
Pumpkin	steam	2½–3	10–16

Dehydration yield of important fruits

Fruit	Yield (%)
Apple	10-15
Apricot	15-19
Banana	14-20
Grape	20-27
Peach	15-19
Pear	15-19

17

Freezing

Freezing is one of the easiest and least time consuming methods of food preservation. The freezing process is a combination of the beneficial effects of low temperatures at which microorganisms cannot grow, chemical reactions are reduced, and cellular metabolic reactions are delayed.

Methods of Freezing

1. Sharp freezing (slew freezing)

In sharp freezing commodities are exposed at -15 to -29^{0}C. It required 3-72 hrs. Here the thawed tissue cannot regain its original water content. The ice crystals formed are large and rupture the cells. The thawed tissue cannot regain its original H_2O content.

2) Quick freezing

It is quick and fast process the food attains the temperature of max ice crystal formation (0 to -4^{0}C) in 30 min or less. Such a speed results in formation of very small ice crystals and hence minimum disturbance of cell structure most foods are quick frozen by one of the following methods.

A) **By direct immersion:** Since liquid are good heat conductors foods can be frozen rapidly by this method in a liquid such as brine or sugar solution at low temperature.

B) **By indirect contact with refrigerant:** It is defined as freezing by contact of the product with a metal surface which is it cooled by freezing brine or other refrigerating media. This is an old method of freezing in which the food or package is kept in contact with the passage through which the refrigerant at - 18 to -46^{0}C flows.

C) **Air blast freezing:** It is a vigorous circulation of cold air to freeze the product. Freezing is done by placing the food stuffs on trays or on a belt and then passed slowly through an insulated tunnel containing air in it. Here the air temperature is approximate -18 to -34^{0}C or even lower. This process is economical and a variety of sizes and shapes can be accommodated.

   I. **Fluidized bed freezing:** This is modification of air blast freezing the foodstuff is fluidized to form a bed of particles and then frozen.

   II. **Plate freezing:** In this method, food products are placed in contact with a cold surface. The cooling temperature of the metal surface is by using cold brine or vaporizing refrigerants. This process is suitable for packaged food products.

3. Cryogenic freezing

In cryogenic freezing commodities are exposed at low temperature (below -60^{0}C). The refrigerants used at present in cryogenic freezing are liquid N_2 and liquid CO_2.

Cryogenic freezing is mostly use in high value commodity such as mushroom, sliced tomato, whole strawberries and raspberries.

4) Dehydro-freezing

This is a process where freezing is followed by partial dehydration. In some fruits and vegetable about 50% of the moisture is removed by dehydration prior to freezing.

5) Freeze - drying

In this process food is first frozen at -18^{0}C on trays in the lower chamber of a freeze drier and then frozen material dried (initially at 30^{0}C for 24 hours and then at 20^{0}C) under high vacuum in the upper chamber. The product is highly hygroscopic, excellent in taste and flavour.

Eg. Mango pulp, orange juice conc, passion fruit juice and guava pulp are dehydrated by this method.

Changes during Freezing of Fruits and Vegetables

1. Physical changes

a) Physical hardening of product

As most of the food have 80 % or more water content which gets converted into ice and ice being a solid form of water though chemically same (H_2O), it gets hardened. Hardening is attributed to the formation of ice crystals that are interwoven with each other very strongly and provide strength and hardness to ice and the product.

b) Expansion during freezing

There is expansion in volume by 6 to 8 % when water is converted into ice. This expansion is due to crystal formation and size of ice crystals is directly related with freezing temperature and rate of freezing. When temperature is very low, rate of freezing or the rate of heat exchange is much faster, the ice crystals formed are smaller in size and fitted more compactly, therefore less expansion in the tissue cells and reverse happens when freezing temperature is high. Ice crystals formed during slow freezing are big and hence expansion in the product is much higher. This expansion may damage tissue cells and distort and deform the product sometimes. Damage in the tissues due to expansion results in higher drip loss or loss of biological exudates, nutrients and weight loss. Maximum water say 80 % of it is converted into ice in the range of 0°C to -5°C.

c) Freezing makes a product lighter

Density of ice (900 kg/m^3) is lesser than pure water (1000 kg/m^3) hence on freezing ice and the product becomes lighter. That is the reason ice floats in the water.

d) Change in the colour of food

Freezing may also affect natural colour of a product due to difference in light diffraction of ice than water, however, this change in colour is reversible when product is thawed. Improper freezing results in active enzymes and oxidative processes. It also denatures pigments and associated proteins. Protiens are part of pigment complexes and they get denatured by cooking, or by acids or by chemical reactions. These changes are permanent and irreversible.

e) Crystallization of water vapors in frozen products

f) Recrystallization of ice inside the product

2. Biological and chemical changes

The biological changes include reduction of micro flora on the surface and interior of food. Freezing has inhibiting effect on the metabolism and reproduction of microbes. As water gets converted into ice and it is not freely available to microbes for their metabolic and physiological activities, they starve and some of them either perish or go into dormant stage. However as and when they get right environment like temperature abuse anywhere during the cold chain, they get active and multiply. Due to this effect reduction in the bacterial load during freezing is witnessed however freezing is not a sterilization process.

a) Protein denaturation

Proteins get denatured due to poor freezing and storage conditions. As proteins bind a lot of water with them, they lose this capacity on their de-naturation hence product loses its juicyness and capacity to swell.

b) Break down of glycogen

c) Loss of electrolytes

Flow chart for freezing of fruits and vegetables

Fruit and vegetables (fully mature)

↓

Washing

↓

Trimming

↓

Cutting into pieces

↓

Scalding / blanching

↓

Keeping syrup/ Brine (Except strawberry which is packed after freezing)

↓

Packing

↓

Freezing.

Freezing point (°F) of important fruits and vegetables

Commodity	**Freezing point (°F)**
Fruits	
Apples	29.3
Apricots	30.1
Blackberries	30.5
Gooseberries	30.0
Raspberries	30.0
Strawberries	30.6
Cherries, sour	29.0
Cherries, sweet	28.8
Nectarines	30.4
Peaches	30.3
Pears	29.2
Plums and prunes	30.5
Quinces	28.4
Vegetables	
Artichokes, Jerusalem	28.0
Asparagus	30.9
Beans green or snap	30.7
Lima Beans	31.0
Broccoli	30.9
Brussels, sprouts	30.5
Cabbage, early	30.4
Cabbage	30.4
Carrots	29.5
Cauliflower	30.6

Contd.

Celeriac	30.3
Celery	31.1
Collards	30.6
Corn, sweet	30.9
Cucumbers	31.1
Eggplant	30.6
Endive and escarole	31.9
Garlic	30.5
Horseradish	28.7
Kale	31.1
Kohlrabi	30.2
Leeks	30.7
Lettuce	31.7
Muskmelons	29.9
Watermelon	31.3
Mushrooms	30.4
Okra	28.7
Onions, green	30.4
Onion, dry	30.6
Parsley	30.0
Parsnips	30.4
Peas	30.9
Peppers	30.7
Potatoes	30.9
Pumpkins	30.5
Radishes	30.7
Rhubarb	30.3
Rutabagas	30.0
Salsify	30.0
Spinach	31.5
Summer squashes	31.1
Winter squashes	30.5
Sweet potatoes	29.7
Tomatoes mature-green	31.0
Tomatoes firm-ripe	31.1
Turnips	30.1
Turnip greens	31.7
Watercress	31.4

18

Fermentation

Fermentation in food processing typically is the conversion of carbohydrates to alcohols and carbon dioxide or organic acids using yeasts, bacteria, or a combination thereof, under anaerobic conditions.

$$\text{Sugar} \xrightarrow{\text{Yeasts}} \text{Alcohol} + CO_2$$

Some aerobic, specific conversions may also be referred as fermentation;

$$\text{Ethylalcohol} + O_2 \xrightarrow[\text{Acetobacter}]{} \text{Acetic acid}$$

Fermentation Methods

A) Acetic acid fermentation

Acetic acid fermentation is produced by the oxidation of ethyl alcohol by bacteria such as *Acetobacter aceti, A. orleansis* and others. The production of vinegar (Acetic acid) from the juice is perhaps one of the oldest organic acid fermentation known.

$$\underset{\text{(Alcohol)}}{C_2H_5OH} + \underset{\text{(Oxygen)}}{O_2} \longrightarrow \underset{\text{(Acetic Acid)}}{CH_3COOH} + H_2O$$

B) Lactic acid fermentation

Lactic acid fermentation is an anaerobic intra-molecular oxidation reduction process. Both homo-fermentative and hetero-fermentative lactic acid bacteria participate in food fermentation.

$$C_6H_{12}O_6 \longrightarrow 2C_3H_6O_3$$
(Dextrose) (Lactic Acid)

C) Alcoholic fermentation

Ethyl alcohol can be produced by fermentation of any CHO containing a fermentable sugar or a polysaccharide that can be hydrolyzed to a fermentable sugar.

$$C_6H_{12}O_6 \xrightarrow{\text{Yeast}} 2C_2H_5OH + CO_2$$
(Dextrose) (Alcohol)

Changes caused in Foods by Fermentation

1. Fermentative changes: Changes that involve carbohydrates.
2. Proteolytic or putrefactive changes: Changes that involve proteins.
3. Lipolytic changes: Changes that involve fats.

In fermentation, fermentative changes are encouraged, whereas other changes are suppressed. Production of acid helps in controlling proteolytic and lipolytic microorganisms.

The End Products of Fermentation Depend upon

(i) The nature of the food,

(ii) The types of microorganisms present,

(iii) The environmental factors affecting microbial activity

The Factors that Affect Fermentation

1. **Acidity:** Fermentative microorganisms are more resistant than proteolytic and lipolytic.
2. **Level of alcohol:** Yeasts cannot tolerate high (12-15%) alcohol levels, (utilized in the production of fermented alcoholic drinks, regular alcohol content of wine 9-13%, fortified up to 20%)
3. **Starter cultures:** Competition for survival, large concentrations dominate the others. In old times inoculation with parts of previous batch at high concentration. Novadays pure starter cultures (concentrates, dehydrated, frozen) available commercially.

4. **Temperature:** Growth rate, optimum growth temperature of specific microorganisms. Optimum temperature regimes are established for various fermentations.
5. **Level of oxygen:** Aerobic, anaerobic microorganisms. Extent of fermentative changes depend on 0_2 concentration. In yeasts growth favored by aerobic, fermentation favored by anaerobic conditions.
6. **Level of salt:**
   I. 10-18% selectively inhibits activity of proteolytic and lipolytic and other spoilage organisms and favors growth of fermentative organisms.
   II. Draws juice out of fruits and vegetables through osmosis.

Fermentation Produces of Fruits and Vegetables

Acetic acid fermentation produce

Vinegar

Vinegar is a product resulting from the conversion of alcohol to acetic acid by acetic acid bacteria, *Acetobacter* spp.

Lactic acid fermentation produce

Sauerkraut

Fresh cabbages are trim clean and shred to 2-5mm thickness. Then 2.25% (w/w) salt is added in the trimmed cabbages and are filled into vats, cover and seal with proper weight and allowed for fermentation. Microorganisms involved in fermentation are started by *L. mesenteroides* and completed by *L. plantarum.*

Pickles

Alcoholic fermentation produce

Wine is an alcoholic beverage made from fermented grapes or other fruits. The natural chemical balance of grapes lets them ferment without the addition of sugars, acids, enzymes, water, or other nutrients.

Crushing the grapes to extract the juice

↓

Alcoholic fermentation

↓

Maltolactic fermentation if desired

↓

Bulk storage and maturation of the wine in a cellar

↓

Clarification

↓

Packaging

Red Grape Wine

Red grape wine is an alcoholic fruit drink of between 10 and 14% alcoholic strength. The colour ranges from a light red to a deep dark red. It is made from the fruit of the grape plant (*Vitis vinifera*). There are many varieties of grape used including Cabernet Sauvignon, Grenache, Nebbiolo, Pinot Noir, and Torrontes. The skins of the grape are allowed to be fermented in red wine production, to allow for the extraction of colour and tannins, which contribute to the flavour. The grapes contribute trace elements of many volatile substances, which give the final product the distinctive fruity character. In addition, they contribute non-volatile compounds (tartaric and malic acids) which impact on flavour and tannins, which give bitterness and astringency.

White Grape Wine

White grape wine is an alcoholic fruit drink of between 10 and 14% alcoholic strength. It is prepared from the fruit of the grape plant (*Vitis vinifera*), and is pale yellow in color. There are many varieties used including Airen, Chardonnay, Palomino, Sauvignon Blanc and Ugni Blanc (Ranken, Kill and Baker, 1997). The main difference between red and white wines is the early removal of grape skins in white wine production. The distinctive flavour of grape wine originates from the grapes as raw material and subsequent processing operations. The grapes contribute trace elements of many volatile substances (mainly terpenes) which give the final product the distinctive fruity character.

Banana Beer

Banana beer is made from bananas, mixed with a cereal flour (often sorghum flour) and fermented to an orange, alcoholic beverage. It is sweet and slightly hazy with a shelf-life of several days under correct storage conditions. There are many variations in how the beer is made. For instance *Urwaga* banana beer in Kenya is made from bananas and sorghum or millet and *Lubisi* is made from bananas and sorghum.

Cashew Wine

Cashew wine is a light yellow alcoholic drink prepared from the fruit of the cashew tree (*Ancardium occidentale*). It contains an alcohol content of between 6 and 12% alcohol.

Tepache

Tepache is a light, refreshing beverage prepared and consumed throughout Mexico. In the past, *tepache* was prepared from maize, but nowadays various fruits such as pineapple, apple and orange are used. The pulp and juice of the fruit are allowed to ferment for one or two days in water with some added brown sugar. The mixture is contained in a lidless wooden barrel called a '*tepachera*', which is covered with cheese cloth. After a day or two, the *tepache* is a sweet and refreshing beverage. If fermentation is allowed to proceed longer, it turns into an alcoholic beverage and later into vinegar. The microorganisms associated with the product include *Bacillus subtilis, B. graveolus* and the yeasts, *Torulopsis insconspicna, Saccharomyces cerevisiae* and *Candida queretana.*

Fortified Grape Wines

Fortified wines are made in the Republic of South Africa and North Africa. Fortified wines are made by adding spirits to wines, either during or after fermentation, with the result that the alcohol content of the wines is raised to around 20 percent, i.e. approximately double that of table wines.

Jack-fruit Wine

Jack-fruit wine is an alcoholic beverage made by ethnic groups in the eastern hilly areas of India. As its name suggests, it is produced from the pulp of jack-fruit (*Artocarpus heterophyllus*). Ripe fruit is peeled and the skin discarded. The seeds are removed and the pulp soaked in water. Using bamboo baskets, the pulp is ground to extract the juice, which is collected in earthenware pots. A little water is added to the pots along with fermented wine inoculum from a previous fermentation. The pots are covered with banana leaves and allowed to ferment at 18 to 30°C for about one week. The liquid is then decanted and drunk. During fermentation, the pH of the wine reaches a value of 3.5 to 3.8, suggesting that an acidic fermentation takes place at the same time as the alcoholic fermentation. Final alcohol content is about 7 to 8% within a fortnight.

Palm Wine

Palm 'wine' is an important alcoholic beverage in West Africa where it is consumed by more than 10 million people. Palm wine can be consumed in a variety of flavours varying from sweet unfermented to sour fermented and vinegary alcoholic drinks. There are many variations and names including *emu* and *ogogoro* in Nigeria and *nsafufuo* in Ghana. It is produced from sugary palm saps. The most frequently tapped palms are raphia palms (*Raphia hookeri* or *R. vinifera*) and the oil palm (*Elaeis guineense*). Palm wine has been found to be nutritious. The fermentation process increases the levels of thiamin, riboflavin, pyridoxin and vitamin $B_{12.}$ Like many African alcoholic beverages, palm wine has a very short shelf-life. The product is not preserved for more than one day. After this time accumulation of an excessive amount of acetic acid makes it unacceptable to consumers. The bark of a tree (*Saccoglottis gabonensis*) may be added as a preservative. The alkaloid and phenolic compounds which are extracted into the wine have antimicrobial effect.

Toddy

Toddy is an alcoholic drink made by the fermentation of the sap from a coconut palm. It is white and sweet with a characteristic flavour. It is between 4 and 6% alcohol and has a shelf life of about 24 hours. It is an important alcoholic beverage in Asia, particularly India and Sri Lanka.

Champagne

This is sparkling wine, produced chiefly in France from grapes.

Pulque

Pulque is the national drink in Mexico, where, it is claimed, it originated with the early Aztecs. Pulque is a traditional beverage that now forms the basis of a national industry, together with the spirits *mezcal* and *tequila* that are obtained from it. Pulque plays an important role in the nutrition of low income people in Mexico with vitamins B being present in nutritionally important levels.

Pulque is a milky, slightly foamy, acidic and somewhat viscous beverage. It is obtained by fermentation of aguamiel, which is the name given to the juices of various cacti, notably *Agave atrovirens* and *A. americana* which are often called the "Century plant" in English. The alcohol content on pulque varies between six and seven percent. The beverage obtained upon distilling pulque is called "*Mezcal*", and if manufactured in the Tequila region from a numbered distillery, it is referred to as "Tequila". The drink is often considered an aphrodisiac. The name *Ticyaol* is given to a good strain that makes one particularly virile. Pulque is frequently the potion of choice used by women during menstruation.

Port

This is a fortified sweet red wine made originally in Portugal, but now produced in some other country also.

Muskat

In the preparation of this wine, especially in Spain, Italy, California and Australia, the Muscat variety of grapes is used.

Tokay

This is the famous fortified wine of Hungary.

Sherry

It is a Spanish wine, matured at high temperature by placing the filled barreles for 3 to 4 months in the open, exposed to the sun where the temperature is as high as 54°C to 60°C.

Cider

Cider is a fermented alcoholic beverage made from apple or pear juice. The alcohol content varies from 2% to 8.5%.

Perry

This is made from pears. Wastes, trimming and cull fruit left over from the canning of pears can also be profitably utilized for making it.

Factors Affecting Wine Fermentation

1. Clarification and pre-treatment of juice

Excessive clarification removes many of the natural yeasts and flora. This is beneficial if a tightly controlled induced fermentation is desired, but less so if the fermentation is a natural one. Long periods of settling out however, encourage the growth of natural flora, which can contribute to the fermentation.

2. Chemical composition of juice

The main consituents of grape juice are glucose (75 to 150 g/l), fructose (75 to 150 g/l), tartaric acid (2 to 10 g/l), malic acid (1 to 8 g/l) and free amino acids (0.2 to 2.5 g/l). The main reaction is the fermentation of glucose and fructose to ethanol and carbon dioxide. However, the presence of nitrogenous and sulphurous products also contributes to the fermentation. The addition of sulphur dioxide to the juice delays the growth of yeast, but does not necessarily inhibit growth of the non-Saccharomyces strains. Fruits generally contain sufficient substrates, soluble sugars - for the yeast to ferment and convert into an acceptable concentration of alcohol. Sugar can be added to fruit juices with a low sugar content, to increase the amount of fermentable substrate.

3. Temperature

Temperature has an impact on the growth and activity of different strains of yeast. At temperatures of 10 to 15° C, the non *Saccharomyces* species have an increased tolerance to alcohol and therefore have the potential to contribute to the fermentation.

4. Influence of other micro-organisms

Other micro-organisms have the potential to influence wine production at all stages of the process. Prior to harvest, yeasts grow on the surface of grapes. Fungicides are used in an attempt to control their growth, but these disturb the natural balance of flora, thus making it difficult to carry out a 'natural' fermentation. Overuse of fungicides can lead to the development of resistant strains of yeast which have the potential to produce toxins which destroy the desirable yeast species. These yeasts are known as 'killer' strains. Other microbes have further chances to influence the fermentation during the clarification process, after fermentation and during maturation and bottling when *Acetobacter* species can oxidise the alcohol and produce acetic acid.

Points to Remember

- In some cases of lactic acid fermentation, yeasts and moulds also participate along with lactic acid bacteria. Lactic acid bacteria are most active at 30^0C temperature.
- Sauer rubem is the fermented product of Turnip.
- Alcohol content of wines ranges from 7-20%.
- Tannin content in Cider is 0.1-0.3%.
- Cider can also prepared from Bael, Jamun, Aonla and Phalsa.
- 'Sirki' a fermented product is prepared from radish tape root.
- 'Khalpi'is prepared by fermentation of cucumber.
- Whisky is a fermented product, which is prepared from malt.
- *Saccharomyces cerevisiae* is known as wine yeast.
- The oriental fermented food 'Poi' is made from taro corms.
- 'Kimchi' is a fermented product from Chinese cabbage.
- Honey wine is known as Mead.
- Distilled grape wine is Brandy.
- Poi product from corm of taro plant.
- Sauer rubben product from turnip.
- Oolong semi-fermented product of tea.

19

Fruit Beverages

Fruit juices which do not undergo alcoholic fermentation are termed as unfermented beverages. They include natural and sweetened juices, RTS, nectar, cordial, squash, crush, syrup, fruit juice concentrate and fruit juice powder. Barley waters and carbonated beverages are also included in this group.

Juice

Juices are of two types:-

1. **Natural Juice (Pure Juice):-** It is the juice, as extracted from ripe fruits, and contains only natural sugars.
2. **Sweetened Juice: -** It is a product which contains at least 85% juice and 10% TSS.

Squash

This is a type of fruit beverage containing at least 25 per cent fruit juice or pulp and 40 to 50 per cent total soluble solids, commercially. It also contains about 1.0per cent acid and 350 ppm sulphur dioxide or 600 ppm sodium benzoate. It is diluted before serving.

Mango, orange and pineapple are used for making squash commercially. It can also be prepared from lemon, bael, papaya, etc. using potassium metabisulphite (KMS) as preservative or from jamun, passion-fruit, peach, plum, raspberry, strawberry, grapefruit, etc. with sodium benzoate as preservative.

Fruit Appetizer/ Spiced Squash

Fruit appetizer is a beverage which essentially consists of moderate quantity of fruit pulp/juice minimum 25% to which can sugar is added for sweetening to raise TSS above 40°Brix along with added salt and spice extract and is consumed after dilution preferably with chilled water as is done in case of squash.

Ready-to-Serve (RTS)

This is a type of fruit beverage which contains at least 10 per cent fruit juice and 10 per cent total soluble solids besides about 0.3 per cent acid. It is not diluted before serving; hence it is known as ready-to-serve (RTS).

Cordial

It is a sparkling, clear, sweetened fruit juice from which pulp and other insoluble substances have been completely removed. It contains at least 25 per cent juice and 30 per cent TSS. It also contains about 1.5 per cent acid and 350 ppm of sulphur dioxide. This is very suitable for blending with wines. Lime and lemon are suitable for making cordial.

Nectar

This type of fruit beverage contains at least 20 per cent fruit juice / pulp and 15 per cent total soluble solids and also about 0.3 per cent acid. It is not diluted before serving.

Crush

It is a beverage similar to squash which is consumed after dilution in water and contains minimum 25% of fruit part and minimum 55% TSS. The acidity is kept at about 1.0-1.2%.

Syrup

It is a beverage similar to squash which is consumed after dilution in water and contains minimum 25% of fruit part and minimum 65% TSS. The acidity is kept at about 1.3%.

Fruit Juice Concentrate

A fruit juice from which water has been mostly removed by heating or freezing is known as concentrate. They contain pure juice with at least 32% TSS.

Fruit Juice Powder

This is the fruit juice which has been converted into a free flowing, highly hygroscopic powder by dehydration or lyophilization e.g. Orange juice powder, lemon juice powder etc.

Carbonated Fruit Beverages

It is the process of dissolving or incorporating carbon dioxide into a ready to serve fruit or non-fruit beverage so that when served it give off gas in fine bubbles and has a characteristic pungent taste. Such beverage containing added dissolved carbon dioxide under pressure is called as carbonated beverage. Fruit juice beverages are generally bottled with CO_2 content varying from 1-8 g per liter.

Barley Water

It is a beverage similar to squash which is consumed after dilution in water and contains minimum 25% of fruit juice and minimum 30% TSS along with 0.25% barley starch. The acidity is kept at 1.0%.

Flowcharts for Processing of Squash

Fruits

↓

Washing

↓

Trimming

↓

Cutting or grating

↓

Juice extraction

↓

Straining

↓

Juice measuring

↓

Preparation of syrup

(Sugar + water + acid, heating just to dissolve)

↓

Straining

↓

Mixing with juice

↓

Addition of preservative

(0.6g KMS or 1.0g Sodium Benzoate/litre squash)

↓

Bottling

↓

Capping

↓

Storage

Processing of RTS Beverages

Fruits

↓

Pulp/juice

↓

Mixing with strained syrup solution

(Sugar + water + acid, heated just to dissolve) according to recipe

↓

Homogenization

↓

Bottling

↓

Crown croking

↓

Pasteurization
(at about 90°C) for 25 minutes

↓

Cooling

↓

Storage

Processing of Cordial

Fruits

↓

Washing

↓

Cutting into halves

↓

Juice extraction

↓

Straining

↓

Addition of preservative
(KMS 1g/litre juice)

↓

Storing in glass container for 10-15 days for clarification
(suspended material settles down)

↓

Symphonizing off the supernatant clear juice

↓

Straining and measuring

↓

Preparation of syrup

↓

Straining mixing of juice with syrup

↓

Addition of preservative
(KMS 0.6g/litre product)

↓

Bottling

↓

Capping

↓

Storing in cool and dry place

Processing of Nectar

Fruits

↓

Washing

↓

Juice/pulp extraction

↓

Straining

↓

Juice measuring

↓

Preparation of syrup
(Sugar + water + acid, heating just to Dissolve)

↓

Straining

↓

Mixing with juice
(Mix citric acid if require)

↓

Addition of preservative
(0.6g KMS or 1.0g Sodium Benzoate/litre Nectar)

↓

Bottling

↓

Capping

↓

Storage

Processing of Barley Water

Barley Flour

↓

Addition of a little water

↓

Making thin paste

↓

Addition of remaining water

↓

Heating to gelatinize

↓

Cooling

↓

Filtering

↓

Addition to juice

↓

Addition of Sugar

↓

Heating just to dissolve sugar

↓

Straining

↓

Addition of essence and preservative
(KMS 0.6g/litre product)

↓

Bottling

↓

Capping

↓

Storage

20

Vinegar

A dilute solution of acetic acid produced by a two-stage fermentation process in the first stage, fermentable sugars is converted into ethanol by the action of yeasts and in the second stage, bacteria of the genus *Acetobacter* oxidize the ethanol to acetic acid.

Vinegar is a liquid produced from the fermentation of ethanol in a process that yields its key ingredient, acetic acid. The acetic acid concentration ranges typically from 4 to 8 percent by volume for table vinegar (typically 5%) and higher concentrations for pickling (up to 18%). Natural vinegars also contain smaller amounts of tartaric acid, citric acid, and other acids.

Steps Involved in Vinegar Production

Two distinct steps are involved its preparation:

1. Conversion of the sugar in fruit, etc., into alchol by yeast (alcoholic fermentation)

The most efficient yeast for fermentation of sugary substances and fruit juices into alcohol are *Saccharomyces ellipsoideus, S. malei* and *S. cerevisiae*. For starchy substances, *S. cerevisiae* is the best. In order to obtain quality vinegar, it is essential to first destroy wild (naturally occurring) yeasts and other microorganisms by pasteurization and then to inoculate pure yeast. The nutrients for the growth of yeast such as phosphates, ammonium and potassium salts and sugars are naturally present in fruit juices and in honey and molasses. The most favourable temperature for the growth of yeast is 25-27^{0}C. Fermentation become abnormal at 38^{0}C and ceases altogether at 41^{0}C and below 7^{0}C.

2. Conversion of alcohol into vinegar by acetic acid bacteria (acetification)

Acetic acid fermentation is brough about by acetic acid bacteria (*Acetobacbacter* spp.) which are strongly aerobic but whose activity is greatly reduced by light. Acetic acid fermentation should, therefore, be carred out in the dark. The nutrients required for bacterial growth are generally present in the alcoholic liquor itself, but in the case of distilled alcohol, malt sprouts, phosphoric acid, potassium carbonate, tri sodium phosphate and ammonium hydroxide are added as nutrients. For acetic acid fermentation, the alcohol content of the fermentation mash is adjusted to 7-8% by dilution with water, because acetic acid bacteria do not grow well at higher concentrations of alcohol. After this adjustment, mother vinegar containing acetic acid bacteria is added at the rate of one part to ten part of fermented mash in order to check the growth of undesirable microorganisms and to hasten the fermentation process.

Vinegar Production

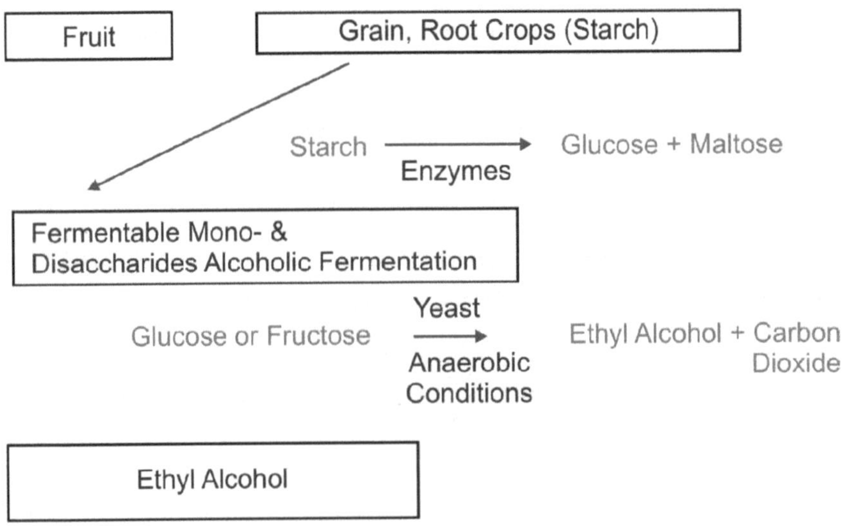

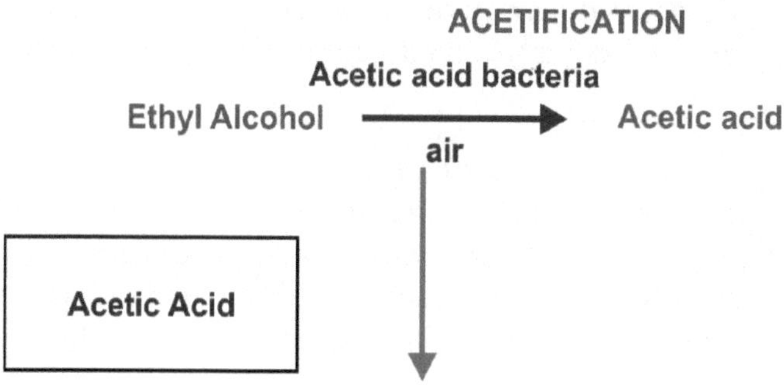

Theoretical conversion

1g glucose → 0.51 g ethyl alcohol → 0.67g acetic acid

Types of Vinegar

Vinegars are of two types

1. Brewed vinegars and
2. Artificial vinegars

1. Brewed vinegar

Brewed vinegars are made from various fruits, starchy material (potato) and sugar containing substances (molasses honey) by alcoholic and subsequent acetic fermentation.

1. **Fruit vinegar:** Generally apple, grape, orange, jamun, peach, pear, pineapple, apricot and banana are used. Vinegar made from apple juice is known as cider or apple vinegar, while that from grapes is known as wine or grape vinegar.
2. **Potato vinegar:** In this case starch is extracted from potato and hydrolyzes by the enzyme diastase before fermentation.
3. **Malt vinegar:** Malt vinegar is derived wholly from malted barley, with or without the addition of the cereal grain, malted or otherwise, the starch of which is saccharified by the diastase of the malt before fermentation. Distilled malt vinegar is prepared by distilling the malt vinegar. The product merely contains the volatile constituent of the vinegar from which it is derived. It is colourless and generally used in the manufacture of pickled onions.

4. **Molasses vinegar:** In this case molasses is diluted to 16% total soluble solid, neutralized whit citric acid and then fermented.
5. **Honey vinegar:** It is prepared from low grade honey.
6. **Spirit vinegar:** Spirit vinegar is the product prepared by acctous fermentation of a distilled alcoholic fluid which in turn is produced by fermentation. It is usually made by alcoholic fermentation of molasses and then distilled prior to acetic fermentation. It is also called as grain vinegar, distilled vinegar, white vinegar or alcohol vinegar.
7. **Spiced vinegar:** Spiced vinegars are prepared by steeping the leaves or spices in an ordinary vinegar.

2. Artificial vinegars

Artificial vinegars are prepared by diluting synthetic acetic acid or glacial acetic acid to legal standard of 4% and are coloured with carmel. Artificial vinegars are also called as synthetic vinegar or non-brewed vinegar.

Three methods used for the production of vinegar are :

1. Slow process
2. Orleans Method
3. Trickling (or quick) Method

Slow process

The slow process, covering a prolonged period, is generally adopted in the country. The juice or sugary solution is filled into barrels and allowed to undergo alcoholic and acetic fermentation slowly. To screen off dust and flies, the bung hole of the barrel is covered with a piece of cloth and the barrel is placed in a damp, but warm place. In about 5-6 months, the sugar solution turns into vinegar. The main drawback of this otherwise simple method is:

I. Alcoholic fermentation is often incomplete
II. Acetic fermentation is very slow
III. Quality of the vinegar is inferior and
IV. Vinegar yield is low.

Orleans method

The earliest vinegar manufacturing method, which was originated in France, was known as the 'Orleans process' or also known as the 'slow process'.

Wooden barrels are used and are filled with alcohol fermenting liquid to almost three quarters full. Holes are drilled into the ends of the barrel a few inches above the liquid level in order to allow air circulation. The holes are left open and netting or screens are placed over the holes in order to prevent insects from getting into the barrels. Fresh vinegar is added to the barrel to acidify the liquid to the point of optimum growth necessary for the vinegar bacteria. The vinegar bacteria settle down into the liquid and a slimy layer develops on the top of the liquid.

This is a slow process because the bacteria come in contact with the air and the substrate only at the surface. The air is supplied through the holes in the barrel. The fermented liquid is allowed to sit for several months with free access to air. Once the alcohol fermenting liquid becomes acetified or is converted to vinegar through activity of organisms growing on or close to the surface of the liquid, a proportion of the vinegar may then be drawn off and replaced with alcohol fermenting liquid and the process is restarted. The acetic acid may become oxidised if the alcohol sources are not constantly added to the vinegar.

The trickling (or quick) method

This process is called the 'Generator' or the 'German' process. Additional supply of oxygen is made available for the bacteria by increasing the surface for the action of the bacteria culture. This increases the rate of fermentation. The equipment used in this process is known as 'Upright Generator'. It has a false bottom and head, ventholes and sparge for discharging the liquor. It consists essentially of three compartments, namely central, distribution and receving compartments.

The generator is in the form of a cylinder, 3.6 to 4.2 metres high and 1.2-1.5 metres in diameter. The details of the three compartments are as follows:

Central compartment: This is filled with beech-wood shavings, corn cobs, pumic stone, rattan shavings or straw to increase the surface area. This chamber is fitted with an adjustable opening near the bottom for admission of air.

Distributing compartment: This is about 30 cm above the central compartment and is separated from it by a partition, which is perforated with a number of small holes. In this compartment, a revolving sprinkler or a tilting trough is fitted to allow the liquid to trickle slowly over the shaving or other material in the central compartment.

Receiving compartment: This is the bottom chamber of the generator in which vinegar is collected. It is separated from the central compartment by a perforated partition about 1.5 metres from the bottom of the generator.

The beech-wood shaving or the filling material is wetted with unpasteurized vinegar. A mixture of two parts of the alcoholic juice and one part of vinegar is fed slowly through the generator to stimulate the growth of vinegar bacteria. Within a few days, the growth of the bacteria becomes sufficient to permit the normal operation of the generator. When the generator is ready, the alcoholic liquid is mixed vinegar in the ratio of 1:2 to increase the acidity from 3 to 3 ½ per cent. It is then passed through the generator. A single passage of this mixture through the generator will convert the remaining alcohol into acetic acid. The process can be made continuous. The progress of oxidation is determined by finding the increase in acidity of the percolating solution.

The generator and the air passage should be cleaned from time to time to eliminate undesirable micro-organism. The optimum temperature at which vinegar bacteria are active is 27^0 to 30^0. In the generator, this is controlled by the adjustment of air intake and by regulating the rate of flow of the liquid. If the temperature rises, the rate of flow of the acidified alcohol is increased and the intake of air reduced. For speedily work in all seasons, generators may be fitted with coils through which cold water or steam may be passed to maintain the optimum temperature.

Alcohol and Acetic Acid Yields

Theoretically, 100 parts of sugar (sucrose or maltose) should yield 53.8 part of alcohol or 70.1 parts of acetic acid. In actual practice, however, even under the most favorable conditions, 44 to 47 parts of alcohol and 50-55 parts of acetic acid only are produced. The loss represents the consumption of sugar in solution by the yeast, loss of alcohol and acetic acid owing to evaporation of oxidation and also loss owing to utilization by vinegar bacteria for their growth. Small quantities of alcohol also may remain unconverted. In order to prepare a vinegar of about 5 per cent acetic acid strength, it is, therefore, necessary, to use a juice with at least 10 per cent sugar (maltose or sucrose) content. After conversion of the whole of the alcohol into acetic acid the vinegar bacteria attack the acid itself. This can be prevented by filling the barrels to the brim and sealing them air tight.

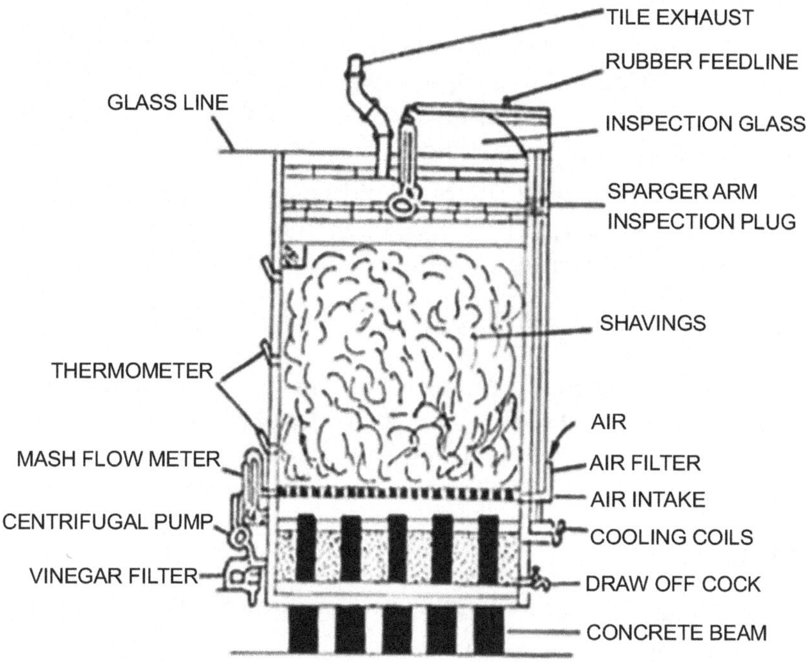

Post-Production Processes

Ageing

To improve its flavour, vinegar is kept in plain oak barrel for about six months. During this period, its harsh flavour changes to a more pleasant aroma and bouquet. This may be owing to the oxidation of vinegar brought about by air entering through the pores in the wood. Acetic acid may also react with alcohol to form ethyl acetate which has a fruit flavour.

Clarification

Before bottling, vinegar must be made sparkling clear. During ageing, most of the suspended material settles down leaving a major portion of the liquid clear. The clear liquid can be siphoned out for further clarification. This can be accomplished either by using finings such as Spanish clay, bentonite, insinglass, casein, gelatin, or by filtering through pulp filters or aluminum plate and frame presses. If finings are used, the vinegar has to be stored for about a month to render it clear.

Pasteurization

The vinegar, after ageing and clarification, is pasteurized to check any spoilage. It is heated in an open vessel to about 66^0C and then cooled to room temperature. It can also be flash-pasteurized by passing it through aluminium pipes surrounded by hot water or steam at 66^0C.

The use of E.K. Seitz filters would ensure the removal of all the micro-organisms present in the vinegar, thereby rendering it sterile. Incidentaly, it also clarifies the vinegars into a sparkling clarity.

Pasteurization with Silver

In this process, the vinegar is either filtered through silver bearing sand, or is electrolysed with silver electrodes at a very low current. The vinegar picks up sufficient concentration of silver ions e.g. about 2 ppm to render it sterile.

Colouring

Caramel colour which is the only permitted colour in the case of vinegar is employed for colouring vinegar.

Problems in Vinegar Making

1. **Wine Flower:** When unfermented juice is exposed to air a film of yiest called wine flower forms on the surface of the liquid. The film causes cloudiness ans also destroys alcohol. The growth of wine flower can be prevented by (1) filling the carboys or barrels up to the brim, (2) addition of 20-25 per cent unpasteurized vinegar and (3) spreading liquid paraffin on the surface of the fermented liquid.
2. **Lactic acid bacteria:** Presence of lactic acid bacteria is very common in fermented juice. These bacteria interfere with acitification, cause cloudiness, produce lactic and other acids which have disagreeable mousy flavour and thus spoil the quality of vinegar. This can be avoided by using 20-25 per cent unpasteurized mother vinegar or a pure culture of yeast.

3. **Insects and worms:** Among insects, vinegar flies (*Drosophila cellaris*), vinegar louse and vinegar mites are important. These can only be avoided by maintain proper samtary conditions. Vinegar (*Anguillula*) are thread-like worms found in vinegar, which destroy the acid and can be killed by heating the vinegar to about 60^0C or removed by filtration. Vinegar eels are strongly aerobic and do not grow if the container is filled upto the brim.

Points to Remember

- pH of vinegar is 3 or below.
- Pickling foods in vinegar is one of the traditional technologies of food preservation.
- The preservative action of vinegar is due to its acetic acid content.
- 0.1% of the undissociated acid will inhibit the growth of most food poisoning and spore-forming bacteria.
- Vinegar can be made from any non-toxic raw material that furnishes a juice or solution containing fermentable sugars.
- The bacteria converting alcohol to acetic acid under natural conditions are film forming organisms on the surface of wine and beer.
- The film was known as 'mother of vinegar' before its bacteriological nature became known.
- Pasteur confirmed that acetic acid is produced only in the presence of the bacteria, but he did not identify them.
- The genus name *Acetobacter* was put forward by Beijerinck in 1900.
- Although *Acetobacter* spp are responsible for vinegar production, pure cultures are hardly used, except in submerged fermentation because of the difficulty of isolating and maintaining the organisms.
- The only member of the genus which is not useful, if not positively harmful in vinegar production is *Acetobacter xylinum* which tends to produce slime
- Recently a new species, *Acetobacter europaeus,* was described.

- Its distinguishing features are its strong tolerance of acetic acid of 4 to 8% in agar, and its absolute requirement of acetic acid for growth.
- 1 gm of alcohol should yield 1.304 gm of acetic acid but this is hardly achieved and only in unusual cases is a yield of 1.1 attained. From the reactions one mole of ethanol will yield one mole of acetic acid and mole of water. It can be calculated that 1 gallon of 12% alcohol will yield 1 gal. of 12.4% acetic acid.
- Over-oxidation can occur and it is undesirable. In over-oxidation acetic acid is converted to CO_2 and H_2O. It occurs when there is a lack or low level of alcohol. It occurs more frequently in submerged fermentations than in the trickle processes.

21

Pickle

The preservation of food in common salt or in vinegar is known as pickling. It is one of the most ancient methods of preserving fruits and vegetables. Pickles are good appetizers and add to the palatability of a meal. They stimulate the flow of gastric juice and thus help in digestion. They are prepared with salt, vinegar, oil or with a mixture of salt, oil, spices and vinegar.

Method of Pickle Preparation

1. **Preservation with salt:** Salt improves the taste and flavour and hardness the tissues of vegetables and controls fermentation. Salt content of 15 per cent or above prevents microbial spoilage. This method of preservation is generally used only for vegetables and some fruits like lime, mango etc. which contain very little sugar and hence sufficient lactic acid cannot be formed by fermentation to act as preservation.
2. **Preservation with vinegar:** The fruits and vegetables preserved in vinegar whose final concentration, in terms of acetic acid in the finished pickle should not be less than 2 per cent.
3. **Preservation with oil:** The fruits and vegetables should be completely immersed in the edible oil. Cauliflower, lime mango and turnip pickles are the most important oil pickles.
4. **Preservation with mixture of salt, oil, spices and vinegar:** The fruits and vegetables can preserve in mixture of salt, oil, spices and vinegar. Cauliflower, carrot, turnip, red chilli, jackfruit

and tomato most important pickle those are prepared by this method.

Problems in Pickle Making

1. **Bitter taste:** Use of strong vinegar or excess spice or prolonged cooking of spices imparts a taste to the pickle.
2. **Blackening:** It is due to the iron in the brine or in the process equipment reacting with the ingredient used in pickling. Certain microorganisms also cause blackening.
3. **Cloudiness:** Cloudiness caused by the use of inferior quality vinegar or chemical reaction between vinegar and minerals.
4. **Dull and faded product:** This is due to use of inferior quality materials or insufficient curing.
5. **Shrivelling:** It occurs when vegetables are placed directly in a very strong solution of salt or sugar or vinegar.
6. **Scum formation:** When vegetables are cured in brine, a white scum always from on the surface due to the growth of wild yeast. This delays the formation of lactic acid and also helps the growth of putrefactive bacteria which cause softness and slipperiness. Addition of 1 per cent acetic acid helps to prevent the growth of wild yeast in brine, without affecting lactic acid formation.
7. **Softness and slipperiness:** This very common problem is due to inadequate covering with brine or the use of weak brine. The problem can be solved by using a brine of proper strength and keeping the pickles well below the surface of the brine.

Flow Chart : Mango Pickle

Mango (peeled and sliced) -1 kg, salt - 200 g, red chilli powder 10 g, asafetida -5 g, fenugreek, black pepper, cardamom (large), cumin and cinnamon (powdered) each 10 g, clove (headless) 6 numbers.

Mangoes
(Mature, green)

↓

Washing

↓

Peeling

↓

Slicing

↓

Putting slices in jar
Sprinkling salt

↓

Putting in sun for one week
(Shaking jar at least twice in a day to mix the salt)

↓

Mixing spices

↓

Storage at ambient temperature
(In cool and dry place)

Flow Chart : Lime Pickle

Select ripe (but not overripe) healthy lime fruits.

↓

Washing

↓

Heat Dip in hot water (60-65^{0}C)
for about five minutes

↓

Cut into four pieces or alternatively cut into smaller, uniform-sized pieces

↓

Brine Ensure that the surface is covered with juice

↓

Dry in the sun for 2-3 days

↓

Mix spices
(To local preference)

↓

Packaging

↓

Storage

↓

(In a cool place, away from sunlight)

Flow Chart : Cucumber Pickle

Cucumber 1 kg, salt 200 g, red chilli powder 15 g, cardamom (large), cumin, black pepper (powdered) each 10 g, clove (headless) 6 numbers, vinegar 750 ml.

Cucumber

↓

Washing

↓

Peeling

↓

Cutting into 5 cm pieces

↓

Mixing with salt

↓

Filling in jar

↓

Standing for 6-8 hours

↓

Draining off water

↓

Adding spices and vinegar

↓

Keeping in sun for a week

↓

Storage
(In a cool place, away from sunlight)

Flow Chart : Green Chilli Pickle

Green chillies - 1 kg, salt - 150 gm, mustard (ground) - 100 gm lime juice - 200 ml (or) amchur - 200 gm, fenugreek, cardamom (large), turmeric, cumin (powdered) each - 15 gm, mustard oil - 400 ml.

Green chillies

↓

Washing

↓

Drying

↓

Making incision

↓

Mixing all spices in a little lime juice

↓

Mixing with chillies

↓

Filling into jar

↓

Adding lime juice and oil

↓

Keeping in sun for a week

↓

Storage
(In a cool place, away from sunlight)

22

Jam, Jelly and Marmalade

Jam is prepared by boiling the fruit pulp with required quantity of sugar to thick consistency, firm enough to hold fruit tissues in position. It should contain not less than 68.5 per cent soluble solids. Jam may be made from a single fruit (apple, strawberry, banana, pineapple etc.) or from combination of two or more fruits. Jam contains 0.5-0.6% acid and invert sugar should not be more than 40%.

- Jam end-point check by Sheet or Flake test (A small portion of jam is taken out during boiling, in a spoon or wooden ladle and cooled slightly, then allowed to drop in water glass. If the product falls off in the form of a sheet or flakes instead of flowing and a continuous stream it means that the end-point has been reached and product is ready, boiling is continued till the sheet is positive.
- End-point for jam making is 105^{0}C, has 68-70%TSS and pH 3.0.
- The jam should be containing 30-50% invert sugar.
- Crystallization of jam occurs due to the less than 30% invert sugar, it can be prevent by adding of corn syrup or glucose along with cane sugar.
- Sticky and gummy jam is due to high percentage of total soluble solid (more sugar). This problem can be solved by addition of pectin or citric acid or both.

- Premature setting is due to low total soluble solid and high pectin content in the jam and it can be prevent by adding more sugar.
- Surface graining and shrinkage is caused by evaporation of moisture during storage of jam, it can be avoid by storing in a cooling place.
- Jam should be stored at 80% humidity for avoid the mould growth.

Jelly

Jelly is a semi-solid product prepared by boiling a clear, strained solution of pectin containing fruit extract with required quantity of sugar and acid. A perfect jelly should be transparent, well set, but not too stiff and should have the original flavour of the fruit. It should be firm enough to retain a sharp edge but should be tender enough to resist the applied pressure. It should not be gummy, sticky or syrupy or have crystallized sugar. Fruits like guava, plum, papaya, gooseberry etc. are used for jelly preparation. Low pectin fruits such as apricot, pineapple, raspberry etc. can be used only after adding small amount of pectin powder.

- Guava is most suitable fruit for jelly making.
- Sour apple, plum, karonda, wood apple, loquat, papaya and goose berry generally used for preparation of jelly.
- Fruits rich in pectin and acid: Sure apple, grape, lemon, orange, jamun, goose berry, cranberry, etc.
- Fruits rich in pectin low acid: Apple, unripe banana, guava, sour cherry, fig, pear, loquat, etc.
- Fruits low in pectin rich acid: Sweet cherry, pineapple, rhubarb, etc.
- Fruits low in pectin and acid: Apricot, peach, pomegranate, raspberry, strawberry, overripe fruits, etc.
- Tartaric acid is best for jelly preparation.
- Pectin is most important constitute of jelly, it is present in fruits form of calcium pectate are responsible for the firmness of fruits.
- Best jelly should have 0.5-1.0% pectin.
- Pectin content determined by alcohol test and jelmeter test.

- The jelly temperature should be 21-38°C at the time of estimation of pectin by jelmeter test.
- Best jelly: pH-3.2, TSS-65% and water 33-38%.
- The acidity of Jelly is 1.0%.
- Methoxyl content of pectin is 7-10%
- Pectin is found in fruits as a Calcium pectate form.
- End-point of jelly judged by Sheet or flake test, drop test (a drop of the concentrated mass is poured into a glass containing water. Settling down of the drop without disintegration denotes the end-point) and temperature test.
- Jelly is ready with 65% TSS at 105^{0}C temperature.
- Jelly is failed to set due to addition of too much sugar, lake of acid the end-point, cooking below the end-point, cooking beyond the end-point and prolonged cooking.
- Cloudy or foggy jelly due to use of non-clarified juice or extract, use of immature fruits, over-cooking, non-removal of scum, faulty pouring and premature gelation.
- Formation of crystals in jelly due to addition of excess sugar and also to over-concentration of jelly.
- Syneresis or weeping of jelly is the phenomenon of spontaneous exudation of fluid from a gel is called syneresis and weeping and is caused by excess of acid, too low concentration of sugar, insufficient pectin, premature gelation and fermentation.
- Jelly pasteurized at 82-85^{0}C for 30 minutes.

Marmalade

Marmalade is a fruit jelly in which slices of the citrus fruit or its peels are suspended. Marmalades are generally made from citrus fruits like oranges and lemons in which shredded peels are suspended.

- Browning during storage of marmalade is very common which can be prevented by addition of 0.09g of KMS per kg of marmalade and not using tin container.

Flow Chart Processing of Fruits Jam

Ripe Firm Fruit

↓

Washing

↓

Peeling

↓

Pulping
(Remove seed and core)

↓

Addition of Sugar
(Add water if necessary)

↓

Boiling
(With continuous starring)

↓

Addition of Citric Acid

↓

Judging of End-Point by Further

↓

Cooking upto 105^0C or 68-70% TSS or by Sheet Test

↓

Filling Hot Into Sterilized Bottles

↓

Cooling

↓

Waxing

↓

Capping

↓

Storage
(At ambient temperature)

Processing of Fruits Jelly

Fruits
(Firm, not over-ripe)

↓

Washing

↓

Cutting into thin slices

↓

Boiling with water
($1^1/_2$ times the weight of fruit for about 20-30 minutes)

↓

Addition of citric acid during boiling
(2g per kg of fruit)

↓

Straining of extract
Pectin test
(for addition of sugar)

↓

Addition of sugar

↓

Boiling

↓

Judging of end-point
(sheet/ drop/ temperature test)

↓

Removal of scum or foam
(one teaspoonful edible oil added for 45 kg sugar)

↓

Colour and remaining citric acid added

↓

Filling hot into clean sterilized bottles

↓

Waxing
(paraffin wax)

↓

Capping

↓

Storage ambient condition

Processing of Marmalade

Ripe Fruits

↓

Washing

↓

Peeling outer yellow portion (Flavedo) thinly

↓

Cutting yellow portion into fine shreds
(1.9-2.5cm long and 0.08-0.12cm thick)

↓

Boiling
(In 2-3 times weight of water for 40 to 60 minutes)

↓

Straining the extract

↓

Testing for pectin content
(Alcohol test)

↓

Adding of sugar
(as required)

↓

Cooking to 103-105^{0}C
(continuous stirring)

↓

Adding of prepared shreds
(Shredded peel boiled for 10-15 minutes in several changes of water for softening and removing bitterness and added @about 62g per kg of extract)

↓

Boiling till jellying point
(continuous stirring)

↓

Testing for end point
(Sheet/drop/temperature test)

↓

Cooking
(to 82-88^{0}C with continues stirring)

↓

Flavoring
(orange oil)

↓

Filling in sterilized bottles

↓

Sealing

↓

Storage at ambient temperature

23

Preserves, Candied, Crystallized and Glazed Fruits and Vegetables

Preserves (*Murabbas*) are prepared from whole fruits and vegetables or their segments by addition of sugar followed by evaporation to a point where microbial spoilage cannot occur. The final soluble solids concentration is reached to about 70 per cent. The finished product can be stored with or without hermetic sealing and refrigeration.

- In preserves fruits can be cooked in syrup by three processes that are: 1) Rapid process, 2) Slow process and 3) Vacuum process.
- Preserves made by vacuum process retain the flavour and colour of the fruit better than other methods.
- Most suitable fruits for preserves making are Aonla, beal, apple, pear, mango, cherry, karonda, strawberry, pineapple, papaya, etc.

Candied Fruits

Fruits/vegetables impregnated with cane sugar or glucose syrup and subsequently drained free of syrup and dried is known as candied fruit/vegetable.

- Most suitable fruits for candied making are Aonla, karonda, pineapple, cherry, papaya, apple, peach and peel of lemon, grapefruit and citron, ginger etc.

- The difference between candied and preserves is that the fruit impregnated with syrup having a higher percentage of sugar or glucose.
- For best candied total sugar content of the impregnated fruit is kept at about 75 per cent to prevent fermentation.

Crystallized Fruits/Vegetables

Candied fruit/ vegetable when covered or coated with crystals of sugar, either by rolling in finely powdered sugar or by allowing sugar crystals to deposit on them from dense syrup are called crystallized fruits.

Glazed Fruits/Vegetables

Covering of candied fruits/vegetables with a thin transparent coating of sugar, which impart them a glossy appearance, is known as glazing.

Problems in Preparation of Preservation and Candied Fruits

1. **Fermentation:** It is due to low concentration of sugar used in the initial stage of preparation of preserves. Sometime fermentation also occurs during storage due to low concentration of sugar and insufficient cooking. This can be prevented by boiling the product at suitable intervals, by adding the required quantity of sugar and by storage in a cool and dry place.
2. **Floating of fruit in jar:** It is mainly due to filling the preserve without cooling and can be avoided by cooling the preserve prior to filling.
3. **Toughening of fruit skin or peel:** It may be due to inadequate blanching or cooking of fruit hence blanching till tender is necessary.
4. **Fruit shrinkage:** Cooking of fruit in heavy syrup greatly reduces absorption of sugar and causes shrinkage. Therefore, fruit should be blanched first or cooked in low-sugar syrup.
5. **Stickiness:** It may develop after drying or during storage due to insufficient consistency of the syrup, poor quality packing and damp storage conditions.

Flow-Sheet for Processing of Preserve and Candy

Mature Fruit
↓
Washing
↓
Preparation for sugar treatment
(according to fruits and vegetables)
↓
Keeping fruit and sugar in alternate layers
(1 kg fruit: 1kg sugar) or steeping fruit in syrup of 40% TSS for a day
↓
Removal of fruit
↓
Increasing consistency of syrup to 60% TSS by boiling
↓
Steeping of fruit
(for a day)
↓
Repeating the process and raising strength of
syrup by 5% TSS to 70% TSS (preserve)
or 75% TSS (candy) on alternate days

Preserve:

↓
Steeping in 70% TSS for a week
↓
Preserve
↓
Draining
↓
Filling in jar or container
↓
Covering fruit with freshly prepared sugar syrup of 68% TSS
↓
Sealing
(airtight)
↓
Storage

Candy:

↓
Steeping in 75% TSS for a week
↓
Draining
↓
Shade drying
(prior to drying, fruit dipped in
boiling water to remove adhering
syrup)
↓
Packing
↓
Storage

❑❑❑

24

Some other Processed Product of Fruit and Vegetable

Food Safety and Standard Authority of India (FSSAI) has adopted the standards of Fruit Product Order (FPO).

Chutney

It is a product prepared by cooking the fruit pulp with added salt, sugar, spices, acetic acid and fruit to a suitable consistency so that it contains TSS not less than 50% and fruit part not less than 40% as per FSSAI specification. Sugar, salt, spices, acetic acid all act as partial preservatives. Preservatives may also be added in chutney for increasing its storability. Generally no artificial colour and flavor are required to be added.

Sauce

Sauce is a product similar to ketchup, prepared from pulps of tomato or other fruit/vegetable, having TSS not less than 15% and cooked to a suitable consistency with added sugar, salt, spices and vinegar. Sugar, salt, spices, acetic acid all act as partial preservatives. According to FSSAI specification fruit sauce should have minimum of 15% TSS and 1.2% acidity. Preservatives and colours may also be added for increasing of appearance and storability.

Tomato Ketchup

Ketchup is a product made by concentrating tomato juice or pulp without seeds and skin, with added spices, salt, vinegar, onion, garlic etc. so that it content not less than 12% tomato solids and generally 28% or more total solid (not less than 25% TSS as per FSSAI specificartion).

Sauce and ketchup are similar products but ketchup is only prepared from tomato while sauce may be made by another fruit and vegetable pulp.

Difference between Ketchup and Sauce

Ketchup	Sauce
1. Prepared from tomato only.	1. Prepared from tomato as well as other fruits and vegetables such as pumpkin, chilli, etc.
2. Minimum TSS is 25%.	2. Minimum TSS is 15%.
3. Minimum acidity is 1.0%.	3. Minimum acidity is 1.2%.
4. Thicker in consistency.	4. Thinner in consistency.
5. Costly.	5. Cheep as compared to ketchup.
6. Only red in colour.	6. May have red, green or other colours.

Tomato Puree and Paste

Tomato pulp devoid of or free from skin and seed, with or without added salt and containing not less than 8.37% salt free from tomato solid is called as *'Medium Tomato Puree'*. It is further concentrated to 12% solid to form *'Heavy Tomato Puree'*.

Tomato paste is concentrated tomato juice or pulp free from skin and seed, containing not less than 25% of tomato solid. It is further concentrated to 33% solid to form *'Concentrated tomato paste'*.

Tomato Soup

Tomato soup is a product having TSS not less than 7°Brix as per FSSAI specification, prepared from tomato juice and which is served hot.

Sauerkraut

Sauerkraut means acid cabbage. It is a clean, wholesome product with a characteristic flavor, obtained by complete fermentation of shredded cabbage in the presence of 2-3% salt. It contains not less than 1.5% acid, expressed as lactic acid. Sauerkraut which has been rebrined in the process of canning or repacking containing not less than 1% acid (lactic acid). Thus it is product of lactic acid bacteria fermentation of cabbage under conditions flavouring the production of lactic acid, acetic acid, alcohol and carbon dioxide.

Fruit Cheese

Fruit cheese is a confection of the type of Karachi Halwa is parepared from fruit like guava, apple, pear and plum. Fruit cheeses have a long shelf-life and are at their best after 3 to 6 months storage. They can be prepared by using fruit pulp, sugar, butter, citric acid, salt and appropriate amount of colour.

Fruit Butter

Fruit butters are prepared from apples, peaches, pears, apricots, plums, grapes, etc., or a combination of these fruits. They have soft butter like consistency can be spread easily on breads. They are made by cooking of pulp with sugar and adding spices. The preparation of fruit butter is similar to that of jam except that fine pulp is used to which small quantites of spices are added.

Fruit Toffee

Pulpy fruit like mango, guava, papaya, etc., can be used for making toffee. It is prepared by using fruit pulp, sugar glucose, skimmed milk powder and appropriate amounts of butter or ghee, essence and colour.

Flow Chart for Processing of Chutney

Fruits

↓

Washing

↓

Sorting

↓

Peeling

↓

Cutting into 4-8 pieces

↓

Heating with some water

↓

↓

Pulping

↓

Pulp (5 kg) → Remove seed, skin, core, etc.

Sugar 2.4-2.6 Kg, salt 140-150g, spices in bag [Red chilli powder (according to taste) 15-25g, black pepper 35-40g, cardamom 70-80g, cumin 40-50g, ginger chopped 350g)] citric acid 15-20 g. →

↓

Cooking over low flame to a thick consistency

↓ → Remove of spice bag

Squeezing the spice bag occasionally

↓

Judging the end point (TSS >50°Brix)

(Added Acetic acid 15-20 ml, Sodium Benzoate 1g)

↓

Mixing

↓

Filling hot into glass jars

↓

Cooling

↓

Labeling and storage

Processing of Tomato Ketchup/Sauce

Tomatoes
(Fully ripe, red)

↓

Washing and Sorting

↓

Cutting and Chopping

↓

Heating at 70-90°C for 3-5 minutes
(To soften)

↓

Straining tomato pulp/juice

↓

Pulping

↓

Cooking pulp with 1/3 quantity of sugar

↓

Putting spice bag

↓

Cooking to 1/3 of original volume of pulp/juice

↓

Removal of spice bag
(After squeezing in pulp)

↓

Cooking and addition of remaining sugar and salt

↓

Judging of end-point
(Tomato solids)

Addition of vinegar/acetic acid and preservative

↓

Filling and cooling

↓

Pasteurization

↓

Crown crocking

↓

Storage at ambient temperature
(In cool and dry place)

Processing of Tomato Puree and Paste

Tomato juice (Strained)

↓

Cooking to desired consistency
(Kettle/vacuum pan)

↓

Judging the end point for puree or pest
(Tomato solids by hand refractometer or reduction in volume by measuring stick)

↓

Filling hot into bottles or cans
(82-88°C)

↓

Pasteurization in boiling water for 20 minutes

↓

Cooling

↓

Storage at ambient temperature

↓

Packaging

↓

Storage

Processing of Tomato Soup

Tomato

↓

Washing

↓

Sorting

↓

Cutting into 4-8 pieces

↓

Heating for 4-5 minute

↓

Pulping

↓

Cooking pulp over gentle flame till 1/3 the original juice volume or up to desired TSS

↓

Putting spice bag

↓

Adding of butter/cream

↓

Cooking to desired consistency with constant stirring

↓

Adding of salt and sugar

↓

Heating to dissolve contents

↓

Removal of spice bag
(After squeezing in pulp)

↓

Filling in can and Sealing

↓

Sterilization at 115°C for 40-45 min.

↓

Cooling

↓

Labeling and Storage

Processing of Sauerkraut

Cabbage
(Mature, white and sound headed)

↓

Removal of core

↓

Trimming
(Removal of outer leaves and defective portions)

↓

Washing

↓

Slicing
(Fine shred)

↓

Putting shreds in jar or vats

↓

Addition of 2.25% salt

↓

Covering mouth of jar with plastic cover

↓

Allowing to ferment for week at 16-22°C

↓

Remove of scum

↓

Drying out juice

↓

Pasteurizing shreds at 75-80°C for 10 minutes

↓

Filling hot into sterilized jars or wide-mouth bottles

↓

Pasteurizing previously drained out juice at 90°C

↓

Pouring hot on to sauerkraut

↓

Sealing

↓

Storage

Processing of Fruit Cheese

Fruit
(Firm and ripe)

↓

Washing

↓

Cutting into thin slices

↓

Boiling with equal quantity of water
(To soften pilp)

↓

Sieving (to remove seed and skin)
and making into fine pulp

↓

Adding sugar, citric acid and butter to pulp

↓

Mixing thoroughly

↓

Cooking till sufficient thick

↓

Adding salt and colour

↓

Removing from fire
(When the mass starts leaving side of pan)

↓

Spreading hot cheese in 0.6 cm thick
layer on tray smeared with butter

↓

Allowing to cool and set

↓

Cutting into small pieces of suitable size

↓

Wrapping in butter paper or polythene sheet

↓

Packaging in dry jars

↓

Sealing

↓

Storage

Processing of Fruit Butter

Fruit
(Firm and ripe)

↓

Washing

↓

Cutting into pieces

↓

Boiling with equal quantity of water
(To soften pulp)

↓

Sieving (to remove seed and skin)
and making into fine pulp

↓

Mixing pulp with sugar

↓

Cooking to consistency of jam

↓

Adding finely ground spices and mixing

↓

Remove from fire

↓

Filling hot into clean sterilization bottles

↓

Sealing

↓

Storage

Processing of Fruit Toffee

Fruit
(Firm and ripe)

↓

Washing

↓

Cutting into pieces

↓

Boiling with equal quantity of water
(To soften pulp)

↓

Sieving (to remove seed and skin)
and making into fine pulp

↓

Concentrating pulp to 1/3 volume by heating

↓

Mixing pulp with sugar and glucose

↓

Putting butter or ghee in pan and placing on fire

↓

Transferring pulp to pan

↓

Cooking till sufficiently solid

↓

Adding milk powder dissolved in little water

↓

Removing from fire

↓

Adding colour and essence

↓

Spreading mass uniformly in 0.5 to 0.75 cm thick layer on a smeared (with butter) tray

↓

Cooling for 4 hours

↓

Cutting into pieces of suitable size

↓

Wrapping in moisture-proof or butter paper

↓

Filling dry jars

↓

Sealing

↓

Storage

❑❑❑

25

Food Quality, Standards and Laws

Quality Standards and Laws

Quality is how well a product or service satisfies the needs of the customer. This includes all aspects related to the needs of the customer such as quality specifications, safetgy, delivery method or date, price etc. Quality can be interpreted in several ways as conformance to the standards, meeting customers' preference/ satisfaction for desired quality attributes, degree of excellence and zero defect products etc. Because of education and consequent greater understanding of implications of poor quality commodities in recent years, consumers have become quality conscious and this fact is also applicable to food and food products. In order to strengthen competitiveness, quality must be incorporated throughout the value added chains right from the harvesting, handling, manufacturing, processing, packaging, storage, marketing and distribution stages, especially in the case of food and food products.

Elements of Food Quality and Safety

The basic functions of a quality control programme are:

- Physical and chemical evaluation of raw materials and processed products
- In-process control of :
  - Raw materials, ingredients and packaging supplies
  - Processing parameters

    - Finished products
- Microbiological analysis and their control in raw materials and finished products
- Control of storage and handling conditions
- Sanitation and waste products control
- Assurance that final products are within the legal and established marketing standards

Steps for Ensuring Food Quality

Quality Control and Quality Assurance are the two steps for ensuring quality. Quality Control is the evaluation of a final product prior to its marketing i.e., it is based on quality checks at the end of production. Quality Assurance is similar to quality control, but has more to do with the process than the product. It is the implementation of quality checks and procedures to immediately correct any failure and mistake that is able to reduce the quality of the interim products at every production step.

The desired high quality of the final product is planned and obtained by conducting Standard Operating Procedures (SOP) that guarantee the desired quality of the interim products at every production step meeting the demands for Good Manufacturing Practices (GMPs). The management approach to long-term success through customer satisfaction is based on the participation of all members of an organization in improving processes, products, services and the working culture and is known as Total Quality Management (TQM). These are the systems that can demonstrate that the organization can meet the specifications and requirements of the customers. They also allow the management of the organization to know that the customer's requirements are being met. Food quality profile for any product is depicted under:

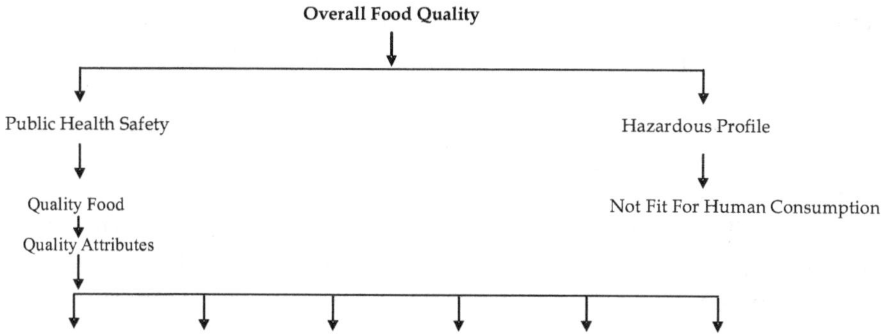

Physical	Chemical	Sensory	Nutritive	Convenience	Microbial
Size	Proximate composition	Appearance	Vitamins	Processing	Total load
Shape	Deteriorative parameters	Colour	Minerals	Shelf life	Yeasts
Viscosity		Smell	Dietary fiber	Packagin	Pathogens
Thickness		Texture	Energy content		Moulds

Good Manufacturing Practices (GMPs) are guidelines to assure that food for human consumption is safe and has been prepared, packed and held under sanitary conditions. These guidelines deal with personnel involved in food processing, building, premises as well as construction and design.

Quality Systems

Food Laws: There are a number of food laws being implemented by various Ministries/Departments. These are primarily meant for two purposes namely, (1) Regulation of Specifications of Food and (2) Regulation of Hygienic Condition of Processing/Manufacturing. Some of these food laws are mandatory and some are voluntary. Food laws are set up and established by authorities as a rule for the measure of quantity, weight, value or quality. Food laws are essential to provide uniform units for weights and measures. The purpose/ benefits of food laws are helpful for farmers and other people engaged in harvesting and food production, those who are engaged in processing and marketing of food, for consumers and government agencies.

Legislations Governing Food Industry in India: With trade liberalization and globalization in the food industry after WTO we have to amend / make changes in our legislations to meet international requirements. In our country, standardization systems fall into two categories. Compulsory legislations are formulated by various Ministries whereas voluntary standards are framed by the organizations with the motto of serving the country. The details of Acts/Orders, their mode of operation, regulations with special features are described in Table 3. Different voluntary legislations are made for the purpose to guarantee stated quality and sales promotion. A number of control orders have been formulated under the provisions of the Essential Commodities Act, which operate on the main objectives of regulating the manufacture, commerce and distribution of essential commodities. There are various commodity boards such as Spices Board, Tea Board, Coffee Board, National Horticulture Board operating in India which undertake research and development work for their respective fields.

International Organizations Governing Food Safety

- World Health Organization (WHO)
- World Trade Organization (WTO)
- Food and Agriculture Organization (FAO)
- Codex Alimentarius Commission (CAC) (Under FAO/ WHO)
- International Organization for Standardization (ISO)
- International Association of Milk, Food and Environmental Sanitarians (IAMFES)
- International Commission for Microbiological Specifications for Foods (ICMSF)
- National Advisory Committee for Microbiological Criteria for Foods (NACMCF)
- International Dairy Federation (IDF)
- Her Majesty's Stationary Office (HMSO)

International Organization for Standardization (ISO)

International Organization for Standardization (ISO) is based in Geneva, Switzerland. Founded in 1947 for the purpose of advancing standardization around the world, this non-government organization is now comprised of over 130 member countries. The ISO 9000 series of quality management standards were developed by the ISO/TC 176 (ISO Technical Committee 176) convened in 1979. It sets out to create a series of internationally recognized quality management standards that represent the essential requirements that every enterprise needs to address to ensure the consistent production and timely delivery of its goods and services to the marketplace. These requirements make up the standards that comprise the quality management system. The ISO 9000 series is able to provide these quality management benefits to any organization of any size, public or private, without dictating how the organization is to be run. The series contains four system standards of varying complexity and completeness which are: ISO 9001, ISO 9002, ISO 9003 and ISO 9004.

The ISO/TC 207 convened in 1993, developed the ISO 14000 series of environmental management standards. The ISO 14000 series of standards represent the essential requirements that every enterprise needs to address in order to control and minimize the impact that its operation, and resulting goods and services, has on the environment.

Codex Alimentarius Commission (CAC)

The Codex Alimentarius Commission (CAC) was created in 1963 by FAO and WHO to develop food standards, guidelines and related texts such as codes of practice under the Joint FAO/WHO Food Standards Programme. The main purpose of this programme is protecting health of the consumers, ensuring fair trade practices in the food trade and promoting coordination of all food standards work undertaken by international government and non-government organizations.

Hazard Analysis and Critical Control Point (HACCP)

Hazard Analysis and Critical Control Point (HACCP) is a food safety programme that was developed nearly 30 years ago for NASA to ensure the safety of food products that were to be used by the astronauts in the space programme. HACCP involves a systems approach for identification of hazards, assessment of chances of occurrence of hazards during each phase, raw material procurement, manufacturing, and distribution, usage of food products and in defining the measures for hazard control.

HACCP is Comprised of Seven Principles

1) Analyze hazards - Potential hazards associated with a food and the measures required to control those hazards are identified that include biological, chemical and physical contaminants.

2) Identify critical control points (CCP). These are points in a food's production at which potential hazards can be controlled or eliminated.

3) Establish preventative measures with critical limits for each control point. These are minimum standards required for the safe preparation of food.

4) Establish procedures to monitor the critical control points. Such procedures include determining how and by whom processing standards are to be monitored.

5) Establish corrective actions to be taken when monitoring has shown that a critical limit has not been met. Therefore, either reprocess or dispose off foods if minimum processing standards have not been met.

6) Establish procedures to verify that the system is working properly for testing and calibrating equipment to ensure their proper functioning which is one typical requirement.

7) Establish effective record keeping in order to document the HACCP system. This would include records of hazards and their control methods, monitoring of safety requirements and corrective actions taken to either prevent problems or how non-conformances are to be prevented from reoccurring.

HACCP enables the producers, processors, distributors, exporters etc., of food products to utilize technical resources efficiently and in a cost effective manner for ensuring food safety. For food industry in India, adoption of HACCP is becoming imperative to reach global standards, demonstrate compliance to regulations/ customer requirements besides providing safer food at all times. HACCP helps in the reduction of contamination, reduction recalling/ product destruction, providing market protection, providing preferred supplier status, demonstrating conformance to international standards, transforming commodities into branded products and facilitating international acceptance.

Standardization Systems for Quality Control of Foods

S.N	Act / Order	Mode of Operation	Regulations	Special Features
I. Compulsory legislations				
1.	Prevention of Food Adulteration (PFA) Act, 1954	i) Ministry of Health & Family Welfare	Makes provisions for prevention of adulteration of food.	Minimum quality standards.
		ii) Directorate General of Health Services	Adulterated, misbranded, and not in accordance with the conditions of license shall be prohibited for selling.	Ensure safety against harmful impurities, adulteration.Mandatory law.
		iii) Central Committee for Food Standards	No such food shall be imported. Standards for the commodities have been specified in the rules. Proprietary foods shall specify the ingredients in the product in the descending order of their composition of the label.	Non-following of PFA Act lead to fine and imprisonment.
2.	Atomic Energy Rules, 1991 (Control of irradiation of food)	Department of Atomic Energy	Regulates the irradiation application in foods. Certificate with the dose and purpose is insisted upon.	Certificate of irradiation indicating the dose and the purpose shall be provided by the competent authority.
3.	Essential Commodity Act, 1954	Ministry of Food	Regulates the manufacture of commodities, commerce and distribution.	Formations of other suborders for easy implementation.

Contd.

4	Fruit Products Order (FPO), 1955	Ministry of Food Processing Industry Central Food Products Advisory Committee	Regulates the manufacture and distribution of all fruit and vegetable products. Exempted from the provisions of the order to products prepared by Drug Control Act and Educational Institutions for training purposes. Quantity shall not exceed 10 kg.License shall be issued after the satisfaction of quality of product, sanitation, personnel hygiene, machinery, equipment and work area requirements as per the schedule specified.	Licensing authority 'FPO' standard mark shall be imprinted on the products
5	Vegetable Oil Products (Regulation) Order, 1998	Ministry of Food and Consumer Affairs	Regulates the production and distribution of all the edible oils. Specifications of the products provided.	Supersedes the Vegetable Oil Products (Control) Order, 1947 and Vegetable Oil Products (Standards of Quality) Order, 1975.BIS Certification for the tin plates used for "*Vanaspati*" packing is deleted.
6.	Sugar (Control) Order, 1966	Ministry of Agriculture and Irrigation Department of Sugar	Regulates the manufacture, quality and sale of sugar.	
7.	Export (Quality Control & Inspection) Act, 1963	Ministry of Commerce	Regulates compulsory, pre-shipment inspection.	AGMARK has been recognized as an agent for inspection and quality control of certain items.

Contd.

		Export Inspection Council 5 Regional Export Inspection Agencies Network of 50 Offices	Exportable commodities list has been notified for pre-shipment inspection. Quality control of various export products is monitored.	Voluntary inspection at the request of foreign buyers and advice of Export Inspection Council is also carried out.
8.	Standards on Weights and Measures Act, 1976	Ministry of Food and Civil Supplies Directorate of Weights and Measures	Prescribed the conditions for packed products with respect to quantity declaration, manufacturing date and sale price.	Providing relief to the weaker sections of society and protecting the consumer in general by guaranteeing the quantity for the amount paid.
9.	The Consumer Protection Act, 1986	Ministry of Food and Civil Supplies	Provision made for the establishment of consumer councils and other authorities for the settlement of consumer disputes.	Protection of the interest of Consumers.
10.	Environ-ment Protection Act, 1986	Ministry of Environment and Forestry	Regulates the manufacture, use and storage of hazardous microorganisms /substances/cells used as foodstuff.	Compulsory for every food plant discharging waste into mainstream to obtain a No Objection Certificate (NOC) from respective State Pollution Board.
11.	The Insecticide Act, 1968	Directorate of Plant Protection, Quarantine and Storage, Ministry of Agriculture	Describes the safe use of insecticides to ensure that residual level doesn't pose any health hazard.	

Contd.

II. Voluntary Standards

1.	Agricultural Produce (Grading & Marketing) Act, 1937	Directorate of Marketing and Inspection	Grade Standards are prescribed for Agricultural and Allied Commodities Grading, sorting as per quality attributes and inspection are included.	Activity is based on marketing and grading at producers' level. Non-following of rules leads to fine and imprisonment. AGMARK Certificate System available.
2.	Bureau of Indian Standards (BIS)	Indian Standards Institution	Prescribing of grade standards, formulation of standards, and specification for foods, prescribing standards for limits of toxic compounds as applicable. Implementation of regulation by promotion through its voluntary and third party certification system, specifying of packaging and labeling requirements.	General cover on hygienic conditions of manufacture, raw material quality and safety are given. Quality and Safety oriented standards.Enforces certification system.
3.	Certification Marks Scheme, BIS Act, 1986 (Rules and Regulations)	Bureau of Indian Standards	Regulates the certification scheme for various processed food products, ingredients and packaging containers.	Ensure the quality to the consumer by certification.

❑❑❑

26

Miscellaneous

New Technique of Food Preservation

1. High Pressure Processing (HPP)

High Pressure Processing (HPP) is a method of food processing where food is subjected to elevated pressures (up to 87,000 pounds per square inch or approximately 6,000 atmospheres), with or without the addition of heat, to achieve microbial inactivation or to alter the food attributes in order to achieve consumer-desired qualities. Pressure inactivates most vegetative bacteria at pressures above 60,000 pounds per square inch. HPP retains food quality, maintains natural freshness, and extends microbiological shelf life. The process is also known as high hydrostatic pressure processing (HHP) and ultra high-pressure processing (UHP).

Advantage of High Pressure Processing (HPP)

HPP at 500-700 MPa allows product modification such as starch gelatinization, gel formation including jams, and tenderization of meat and fish. The precise uses of HPP depend on the pressures and temperature selected.

HPP pasteurization at room temperature and 200-600 MPa, reduces the number of (vegetative) micro-organisms and prevents product spoilage. Treatment times can vary from 1-10 minutes and throughput is 1-2 tones per hour. Products can last up to one month in the refrigerator, but additional preservation is necessary for storage at room temperature.

For HP sterilization, products are heated to 70-90°C at 800 MPa, and adiabatic heating - without heat transfer, means temperatures rise to 120 °C. Treatment times are generally less than three minutes and shelf-life is extended to months at room temperature.

Applications of HPP

Like any other processing method, HPP cannot be universally applied to all types of foods. HPP can be used to process both liquid and solid foods. Foods with a high acid content are particularly good candidates for HPP technology. Some products that are commercially produced using HPP are cooked ready-to-eat meats, avocado products (guacamole), tomato salsa, applesauce, orange juice, and oysters.

HPP cannot yet be used to make shelf-stable versions of low-acid products such as vegetables, milk, or soups because of the inability of this process to destroy spores without added heat. It can be used to extend the refrigerated shelf life of these products and to eliminate the risk of various food-borne pathogens such as *Escherichia coli*, *Salmonella* and *Listeria*. Another limitation is that the food must contain water and not have internal air pockets. Food materials containing entrapped air such as strawberries or marshmallows would be crushed under high pressure treatment, and dry solids do not have sufficient moisture to make HPP effective for microbial destruction.

HPP processed foods stored

HPP products currently marketed worldwide are primarily distributed refrigerated. In some cases this is necessary for safety (to prevent the growth of spores in low-acid foods). For acid foods, refrigeration is not a necessity for microbial stability, but is used to preserve flavor quality for extended periods of time.

2. Combined Method of Preservation (Hurdle Technology)

A judicious combination of more than one method mentioned above for synergistic preservation (hurdle technology).

The trend of using a wide range of mild preservation techniques has emerged to be known as combined preservation or barrier (Hurdle) technology. Hurdle in food is defined as the substance or the processing step or various preservation factors, inhibiting the growth of various microorganisms resulting in the death of microorganisms. It advocates the deliberate combination of existing and novel preservation techniques in order to establish a series of preservative factors (hurdles) that any microorganisms present should not be able to overcome. It requires a certain amount of effort from a microorganism to overcome each hurdle. Higher the hurdle, greater the effect. The quantification of various factors in terms of pH, redox-potential, water activity, temperature, preservatives etc. gives the successful level of combination of hurdles, which lead to failure of homeostasis. As a result the microorganisms will not multiply i.e. either remain in lag phase or die. Several tropical and semitropical fruits and vegetables like carrot, capsicum and coconut are processed by hurdle technique by slight reduction of water activity (a_w = 0.92-0.95), lowering of pH (below 4.5) and mild heat treatment (in-pack pasteurization at 85^{0}C) or treatment with antimicrobial additives with a view to control microbial growth, packed in flexible polymeric pouches and were evaluated for their shelf stability under ambient conditions.

Potential hurdles for use in preservation of food

a. **Physical hurdles:** High temperature (sterilization, pasteurization and blanching), low temperature (chilling and freezing), electromagnetic radiation, packaging film, modified atmosphere packaging (gas, vacuum, moderate vacuum and active packaging), aseptic packaging etc.

b. **Physico-chemical hurdles:** Low water activity (a_w), low pH, low redox-potential (E_h), common salt (NaCl), nitrate, CO_2, O_2, O_3, organic acids, lactic acid, lactate, acetic acid, acetate, ascorbic acid, sulphite, smoking, phosphates, glucono-o-lactone, phenols, chelators, ethanol, propylene glycol, maillard reaction products, spices, herbs, lactoperoxidase and lysozyme.

c. **Microbially derived hurdles:** Competitive flora, protective cultures, bacteriocins and antibiotics.

Advantages of Hurdle Technology

The technology leads to the development of high quality food that is shelf stable, with superior quality and with fresh like characters, further more this approach is not single-targeted but multi-targeted. There is every possibility that different hurdles will have an additive or synergistic effect in food. The concept of hurdle technology has proved extensively useful in optimization of traditional foods as well as development of novel products. For securing stable, safe and tasty foods, linkage between hurdle technology (used for food design), the HACCP concept (used for process control) and predictive microbiology (used for process refinement and food safety) is indispensable.

Several methods such as freezing, canning, dehydration, chemical preservation etc. are commonly used for preserving foods. However, all these processes are based on a relatively few parameters or 'hurdles', combination of which decisively govern microbial stability and nutritional quality of almost all foods.

3. Microwave Heating

Microwave heating refers to the use of electromagnetic waves of certain frequencies to generate heat in material. When a microwavable container with food is placed in a microwave oven and then an oven is activated, the food at the edge of the container heats faster and a temperature gradient develops between the centre and the edges. For food applications most commonly used microwave frequencies are 2450MHz and 915MHz.

Materials for Microwave Heating

Plastic containers are commonly used for microwave cooking. High density polyethylene can be used for foods with high water content, it cannot be used for foods with high fat or high sugar content as these foods may reach temperature above 100°C during microwave cooking. Paper and board can also absorb some microwave energy. It is not ideal for microwaved food because the strength of the paper would be affected when wet and not all types of paper are suitable for microwave cooking. When food is microwaved, heat is also retained in the glass. The degree of energy absorption depends on the types of glass. Moreover, microwave energy can be superimposed at the centre after passing through the glass containers, particularly the ones with small radius.

Applications of Microwave Heating

Microwave is used in baking, concentration, cooking, curing, drying, finish drying, freeze drying, pasteurising, sterilizing, tempering and thawing

4. Ohmic Heating

Ohmic heating is based on the principle of passing an electric current through an electrically conducting product, as with microwave energy, electric energy is transformed into heat. Ohmic heating is an efficient way of processing foods containing large solid particulates unlike conventional method of processes such as canning and aseptic processing. After heating products can be cooled in continuous heat exchangers and then aseptically filled into pasteurized containers in a manner similar to conventional aseptic packaging. Both high and low acid products can be processed by this method. Ohmic heating cannot be used for all the food products. It is depending upon the product's electrical conductivity and on whether the product is an insulator or a conductor. In this process simultaneous and uniform heating of solid and liquid phases can be achieved, thus reducing the danger of under processing as well as nutritional loss. The critical parameters affecting ohmic heating include the electrical conductivity of the food, temperature dependence of electrical conductivity, the design of heating device, the residence time, distribution time, thermo physical properties of foods and electric field strength.

Fruits were analysed for Electrical conductivity during Ohmic heating. At 25°C the electrical conductivity of pineapple was very low and significantly different than apples and pear. Electrical conductivity of peach and strawberry was high. At higher temperatures (40-140°C), apples and pineapple had low conductivity. Strawberry and peach had higher conductivity and significantly different compared to other fruits. The gap in the electrical conductivity between strawberry and peach, and other fruits increased with the temperature. Cloudberry jam was treated with ohmic heating and also by traditional method. There is no significance difference for sensory attributes and rhelogical properties between ohmic heated jam and jam prepared by another method.

Applications of Ohmic Heating

Used in blanching, evaporation, dehydration, fermentation, extraction and value added process.

5. Pulsed Electric Fields (PEF)

Pulsed electric fields (PEF) is a non-thermal method of food preservation that uses short pulses of electricity for microbial inactivation and causes minimal detrimental effect on food quality attributes. PEF technology aims to offer consumers high-quality foods. For food quality attributes, PEF technology is considered superior to traditional thermal processing methods because it avoids or greatly reduces detrimental changes in the sensory and physical properties of foods.

The Principles of Pulsed Electric Field

The basic principle of the PEF technology is the application of short pulses of high electric fields with duration of microsecond's micro- to milliseconds and intensity in the order of 10- 80 kV/cm. The processing time is calculated by multiplying the number of pulses times with effective pulse duration. The process is based on pulsed electrical currents delivered to a product placed between a set of electrodes; the distance between electrodes is termed as the treatment gap of the PEF chamber. The applied high voltage results in an electric field that causes microbial inactivation. The electric field may be applied in the form of exponentially decaying, square wave, bipolar, or oscillatory pulses and at ambient, sub-ambient, or slightly above-ambient temperature. After the treatment, the food is packaged aseptically and stored under refrigeration. Applied to a food product held between two electrodes inside a chamber, usually at room temperature. Food is capable of transferring electricity because of the presence of several ions, giving the product in question a certain degree of electrical conductivity. So, when an electrical field is applied, electrical current flows into the liquid food and is transferred to each point in the liquid because of the charged molecules present.

Advantage of Pulsed Electric Fields (PEF)

PEF inactivates vegetative micro-organisms including yeasts, spoilage micro-organisms and pathogens, and it can be used to pasteurize fluids such as juices, milk and soups without using additives. This technology can substitute for conventional heat pasteurization or it can operate at room temperature to retain quality and heat-sensitive vitamins.

PEF can be used as continuous process but, after processing, products have to be packaged hygienically and kept cool during storage. PEF causes the formation of large, permanent pores in cellular tissues, which can be used to improve juice yield, increase concentrations of functional components and enhance the characteristics of dried produce. It can support or replace conventional processing techniques such as enzymatic maceration and mechanical disintegration. However, low electric field strength and/ or pulse number causes reversible cell rupture stimulating a stress reaction in plants or cell cultures and allowing enzymes or proteins with potential health benefits to be harvested.

Storage of PEF Processed Food

PEF pasteurized products currently are stored refrigerated. In some cases (for example, milk), this is necessary for safety (to prevent the growth of spores in low-acid foods). For acid foods, refrigeration is not necessary for microbial stability, but is used to preserve flavor quality for extended periods of time.

6. Pulsed Light or High Intensity Light Technology

High intensity light is also described as pulsed broad spectrum white light is a decontamination or sterilization technology that can be used for the rapid inactivation of micro organisms on food surfaces, equipments and food packaging materials. Surface decontamination of food products using pulsed high intensity light has many potential benefits to the food industry. It is a non thermal preservation intervention, with the ability to minimize the deleterious effects of thermal processing and chemical treatments on quality and sensory attributes.

The food preservation that involves the use of intense and short duration pulses of broad-spectrum "white light". The spectrum of light for pulsed light treatment includes wavelengths in the ultraviolet (UV) to the near infrared region. The material to be treated is exposed to a least 1 pulse of light having an energy density in the range of about 0.01 to 50 J/cm^2 at the surface. A wavelength distribution such that at least 70% of the electromagnetic energy is within the range from 170 to 2600nm is used. The material to be sterilized is exposed to at least 1 pulse of light (typically 1 to 20 flashes per s) with a duration range from 1 μs to 0.1 s. For most applications, a few flashes applied in a fraction of a s provide a high level of microbial inactivation.

Applications of High Intensity Light Technology

Used in decontamination of vegetables, dairy products. Also used in microbial inactivation of water, sanitation of packaging materials and disinfection of equipment surfaces.

7. Pulsed X-Rays

Pulsed X-ray is a technology that utilizes a solid state opening switch to generate electron beam X-ray pulses of high intensity. Electrons have limited penetration depth of about 5cm in food while X-ray have significantly higher penetration depths (60-400cm) depending upon the energy used.

Applications of Pulsed X-Rays

Use in examination of packaged food for tramp materials, inactivation of *e-coli* in meat also used to eliminate *Salmonella* from foods.

8. Ultrasonics

The frequency of sound waves audible to human ear ranges from 20Hz to 20 kHz. The sound waves having frequencies greater than 20 kHz are called "Ultrasonics" or "Supersonics". The term supersonics is used for sound waves having velocities greater than that of sound. Sound waves of frequencies less than 20 Hz are called "Infrasonics".

Ultrasonically enhanced drying was carried out at lower temperatures than the conventional methodology which reduces the probability of oxidation or degradation in the material. By employing ultrasound the heat transfer between a solid heated surface and a liquid is increased by approximately 30-60%. Application of power ultrasound is benefit in ice cream manufacturing by reducing crystal size, preventing incrustation on freezing surface. Inconsistency in beef tenderness has been rated as one of the major problems faced by the meat industry. Tenderness is influenced by composition, structural organization and the integrity of skeletal muscle. Ultrasound-assisted process of meat tumbling caused the significant improvement of the yield, tenderness and juiciness of the end product. Sonication resulted in decreased drip loss and shear force of PSE meat.

Applications of Ultrasonics

Used in pasteurization at mild heat, extraction, enzyme inactivation, emulsification, crystallisation, viscosity alteration and degassing, spraying or coating, anti fouling and de-foaming.

Functions of Food

No one live without food beyond a short period. Food is therefore necessary for all living beings.

Food performs the following four functions:

Energy

Food supplies the fuel or energy needed to perform the many tasks of everyday living. Human need energy to think, breathe, walk, sit, speak and even sleep. We get energy from carbohydrates, proteins and fats, all are supplied from food.

Growth and Repair

Food provides the materials needed to build, repair and maintain body tissues. Proteins, fats and minerals are the best nutrients for growth. Growing bodies need extra amounts of these nutrients. Every person, whether growing or not, is going through a continual repair process of replacing injured or dead cells. It is food that supplies the nutrients necessary for this process.

Regulation

Food supplies the substances that help regulate the body's processes. Water, vitamins and minerals help regulate breathing, the nervous system, digestion, blood circulation and the elimination of waste products from the body. They help keep all the systems in the body working properly.

Protection

Vitamins, minerals and protein keep the body's tissues and organs healthy. Healthy organs are less likely to be attacked by disease.

Cold Chain

A cold chain is a temperature-controlled supply chain. Perishable goods like fresh fruits, vegetables and flowers, follow a cold chain while in transit, in warehouses, and while waiting to be loaded onto transport.

The Importance of the Cold Chain

Maintenance of the cold chain is the best way to maintain the quality of a product and minimise all forms of deterioration after harvesting, including:

- Weight loss resulting in wilting and limpness
- Softening
- Bruising
- Unwanted ripening
- Colour changes
- Texture degradation and
- Development of rot and moulds.

The export of fresh produce often involves long journey times and frequent handling. This makes effective cold chain management more difficult, but even more essential, to ensure that the product finally sold retains maximum freshness.

Cold chain is most important for handling of fruit and vegetables. Maintenance of low temperature at different stage of handling helps in reduction of losses and retention of quality. Suggested cold chain system for handling of fruits and vegetables from orchard to consumer are as under:

Fruit/Vegetables

↓

Harvesting at right stage of maturity

↓

Sorting

↓

Transporting to packing house

↓

Washing/cleaning

↓

Precooling and packing

↓

Transporting
(Through reefer vans/containers/CA containers)

↓

Cold storage

↓

Marketing through air - cooled kiosks

Papain

Milky latex is the major constituent of immature papaya fruits. Papain is an enzyme prepared from dried latex if unripe papaya is in great demand in the international markets.

Processing Flow Chart for Manufacturing of Papain from Papaya Fruits

Papaya Fruits
(Fully mature but not ripe, about 90-100 day old)

↓

Making Incision
(Longitudinal and four side of fruits from
stalk end to the tip in the morning, depth about 0.3 cm)

↓

Collecting of Latex

↓

Filtration

↓

Mixing with 1000 ppm Sulphur Dioxide

↓

Sun Drying (Up to 5% moisture content)

↓

Powdering

↓

Blending

↓

Packaging

↓

Storage

Browning of Food

Browning recation occur very widely in food materials. This colours produced range from pale yellow to dark brown or black, depanding on the type of product and the extent of the reaction. In some foods browning is considered desirable, e.g., honey, chocolate, brown crust of baked product, etc., while in other food it is detrimental, as in darkening of dehydrated fruits, vegetables, etc.

The term "browning" in relation to food refers to several different processes which can be divided into enzymatic and non-enzymatic browning reactions.

Type of Browning

1. Enzymatic browning

Enzymatic browning is a less desirable series of reactions. It occurs in certain fruits and vegetables when phenolic compounds react with oxygen in the air, which results in brown complexes. No or few flavours are produced. Examples are the browning of cut apples and bananas and the browning of tea (from green to black tea).

2. Non-Enzymatic browning

Maillard reaction

Maillard reaction is a non-enzymatic browning reaction, caused by the condensation of an amino group and a reducing compound, resulting complex changes in biological and food system. This reaction was described for the first time by Louis Maillard in 1912. Maillard reaction occurs when virtually all foods are heated, and also occurs during storage. Most of the effect of Maillard reaction, including the caramel aromas and golden brown colors, are desirable. Nevertheless, some of the effect of Maillard reaction, including foods darkness and off-flavor development, are undesirable.

Reducing Sugar + Amino Acid ⟶ Brown Pigments + Flavours

Caramelization

The second type of non-enzymatic browning is **caramelization** that occurs when carbohydrates/sugars in any food are heated. Caramelization results in light to dark brown and new flavours in the product. Caramelization plays an important role in roasting of coffee and commercial caramels are added as food colours or flavours.

High temperature

Sugar ⟶ Brown Pigments + Flavours

Ascorbic Acid Browning

Ascorbic acid browning is the spontaneous thermal decomposition of ascorbic acid under both aerobic and anaerobic condition and either in the presence or absence of amino-compounds.

Water Activity (a_w)

Water activity (a_w) or equilibrium relative humidity (ERH) quantifies the active part of the moisture content or 'free water' as opposed to the total moisture content which also includes 'bound water'. Water activity (a_w or ERH) not only determines the stability of moisture content (and of the parameters it influences) but is also a decisive factor in various problems where moisture content is of less interest.

The moisture content of a product is usually defined as the percent weight of water content in relation to the dry weight of the sample.

Water activity or equilibrium relative humidity measures the water-vapour pressure generated by the water present in a hygroscopic product.

Aw = p/ps and ERH = 100 x a_w where,

p = water-vapour pressure above the product surface

ps = water-vapour pressure above the surface of pure water at the product temperature

On the basis of moisture content, the food products can be Broadly put into three categories as:

- **Low moisture foods** e.g., dried or freeze dried foods having moisture content of 5-15 per cent.
- **Intermediate moisture foods** e.g., cakes and dates with a moisture content of 20-40 per cent.
- **High moisture foods** e.g. fresh fruits and vegetables of greater than 40 per cent of moisture.

Water activity is a critical factor that determines shelf life. Most bacteria, for example, do not grow at water activities below 0.91, including pathogens such as *Clostridium botulinum.* Below 0.80 most molds cannot be grown and below 0.60 no microbiological growth is possible. However, there remain a number of food spoilage microbes that can grow within the range 0.8 - 0.6. The risk of food poisoning must be considered in low acid foods (pH > 4.5) with a water activity greater than aw 0.86. *Staphylococcus aureus,* a common food poisoning organism, can grow down to this relatively low water activity level. Foods that may support the growth of this bacterium include cheese and fermented sausages stored above correct refrigeration temperatures.

Microorganisms grow at this a_w and above	**Food examples**
0.95 Pseudomonas, *Escherichia, Proteus, Shigella, Klebsiella, Bacillus, Clostridium perfringens,* some yeasts	Highly perishable foods (fresh and canned fruits, vegetables, meat, fish), milk, cooked sausages, breads, foods with up to 4 oz sucrose or 7% NaCl
0.91 Salmonella, Vibrio parabaemolyticus, C. botulinum, Lactobacillus, some molds	Some cheese (Cheddar, Swiss, Provolone), cured meat, fruit juice concentrates with 55% sucrose or 12% NaCl
0.87 Many yeasts, *Candida, Torulopsis, Hansenula micrococcus*	Fermented sausage, sponge cakes, dry cheese, margarine, foods with 65% sucrose or 15% NaCl
0.80 Most molds, most *Saccharomyces* spp., *Debaryomyces, Staphylococcus aureus*	Most fruit juice concentrates, condensed milk, syrup, flour, high-sugar cakes, pulses containing 15-17% moisture
0.75 Most halophilic bacteria, *Mycotoxigenicaspergilli*	Jam, marmalade, glace fruits, marzipan, marshmallows
0.60 Osmophilic yeasts, few molds	Dried fruits with 15-20% moisture, caramel, toffee, honey
0.50	Noodles with 12% moisture, spices with 10% moisture
0.40	Whole egg powder with 5% moisture
0.03	Whole milk powder with 2-3% moisture, dehydrated soups

Source: Water Activity of Some Foods and Susceptibility to Spoilage by Microorganisms Adapted from Beuchat (1981).

Tetra-Packaging

Tetra Pak aseptic cartons are made of three basic materials that together result in every efficient, safe and light-weight package. Each material provides a specific function:

1. Polyethylene: Protects against moisture
2. Paper: For stability & strength

3. Polyethylene: Adhesion
4. Aluminium: Barrier to oxygen, flavour & light
5. Polyethylene: Adhesion
6. Polyethylene: Seals in the liquid

It is used to store the fruit beverages and RTS beverages.

- Paper (80%): to provide strength and stiffness
- Polyethylene (15%): to make packages liquid tight and to provide a barrier to micro -organisms
- Aluminium foils (5%): to keep out air, light, and off-flavours-all the things that can cause food to deteriorate.

Combining each of these three materials has enabled Tetra Pak to produce a packaging material with optimal properties and excellent performance characteristics.

- Higher degree of safety, hygiene and nutrient retention in food
- Preserving taste and freshness
- Can be kept for months with no need for refrigeration or preservatives
- Efficient (a filled package weight is 97% product and only 3% packaging material), using a minimum quantity of materials necessary to achieve a given function
- A good example of resource efficiency is its light-weight (among the lightest packages available)

27

Queries

- 'Prunes' is prepared from peach.
- "Arrak" liquor is prepared from date palm.
- "Hayane" variety of date palm suitable for juice making.
- "Kanji" is made from black carrot.
- "Kirschwascer" a wine prepared from cherry.
- "Shahi" variety of Litchi suitable for canning.
- 0.32-0.47 head space needed to be leaving inside the cans when cans are closed on double seaming machine.
- 1 rad= 10^{-2} J Kg^{-1} 1 Gy=100 rads.
- 1 ton refrigeration=200 Btu/min = 3.517 kJ/s = 3.517 KW.
- 1 tonne of refrigeration is about 10% larger than 1 ton of refrigeration.
- 1 tonne of refrigeration= 13,898 kJ/h=3.861 KW.
- 1:6 ratios of protein and carbohydrate in balance diet are recommended.
- 100 g rose hip syrup is containing 150 mg ascorbic acid.
- 100°C Salometer reading = 26.5% salt concentration (maximum solubility of salt in water is 26.5%).
- 10-100 ppm is the optimum ethylene concentration for ripening of fruit.

- 1Gy=1 J of energy per kilogram of food
- 20°C is optimum temperature for ripening of most of fruits.
- 5.5-7.0 most favorable pH ranges for growth of bacteria.
- 8-HQC shows toxicity in Daffodil.
- A blue-black colour formation on the interaction of food item with Iodine solution is an indication of the presence of starch.
- A process in which the flowers are kept loosely in a big container so that air can circulate around stem is known as conditioning/hardening.
- A storage disorder in apple, which occurs due to accumulation of toxic levels of volatiles in skin, is called apple scald.
- A very minute leak in can that only permits air to move but does not necessarily allow micro-organism to enter is called breather.
- A winery by-product of grape, which settles on walls and base of wine tanks, is called argol.
- ABA is strongest senescence stimulator.
- Abscissic acid having significant role in regulation of senescence in flowers.
- Absolute humidity= Mass of water vapour / Unit mass of dry air (kg/kg).
- Accelerated and Vibration tests are used for test the non-mechanical damage in packaging.
- According to FSSAI specification, 45% fruit content and 65°Brix TSS require for jelly.
- According to FSSAI specification, the food products without preservation are pickles and chutneys.
- Acetic acid preservative is added in tomato products.
- Acid resistant lacquers are ordinary gold coloured enamel and can when treated with it are called R-enamel canes and used for packaging fruit of acid group with soluble colouring matter.
- Aflatoxin are produced by *Aspergillus flavus* while Patulin by *Penicillium* sp.
- Air cooled storage is useful to store commodities in higher elevation zones.

- Ajeone is sulfur component that occurring "Natriceutical" that reduce the risk of stroke and heart diseases.
- Ammonia based refrigerant is commonly used in cold storage in our country.
- Amongst all the hydrocarbons, ethylene is the most effective fruit ripening agent because it possesses short chain length and double bond.
- Anthocyanin is responsible for red coloure in Roses.
- Anti-oxidants for the prevention of oxidation of fact and prevent ageing and diseases.
- Antioxidants soluble in water are known as hydrophilic.
- Antioxidants soluble in fat are known as hydrophobic.
- Aphla, beat and gamma rays have <100 nm wave length.
- As per FSSAI specification tomato juice should not less than 5% total solid.
- Aseptic canning concept is given by Martin in 1950.
- *Bacillus stearothermophillus* and *Clostridium thermo-saccharolyticum* are thermo-philic bacteria.
- Baking soda use as a leavening agent in food industry.
- Balanced diet is the food that supplies all the essential nutrients in required quantity.
- Banana is cured in smoke for ripening (JRF-2011).
- Benzylaminopurine delay chlorophyll degradation and senescence of leafy vegetables by maintaining high protein level on the applied tissues.
- Best degreening temperature 27°C, 85-90% relative humidity and 20 ppm ethylene concentration for citrus.
- Beta carotene is known as "Natures most potent naturally occurring antioxidant".
- Bitter principle of mustard and horse radishes is *sinnigrin.*
- Bitter taste in brinjal fruit is due to solasodine.
- Bittering component mostly formed as a result of heat treatment of juice during processing is limonin.
- Bitterness in grape fruit juice is due to naringin.

- Black spot is reduced by washing radishes in Chlorinated water.
- Blanched vegetables are sulphited by placing them in 0.5% solution of potassium metabisulphite for sufficient time is called sulphitation.
- Bromelin is present in Pineapple.
- C. Funk coined the term vitamin.
- C_6H_5COOH is a formula of Benzoic acid.
- CAC is a subsidiary body of FAO & WHO.
- CaC_2 is used for dehydration of flower.
- $CaCl_2$ is useful in increasing firmness of fruits.
- $CaCl_2$ used in water to raise the temperature up to 121.1^{O}C.
- Can with a mild positive pressure is called flipper.
- Card board boxes most suitable packaging material for cut flower packaging.
- Cauliflower having maximum post harvest losses.
- Caustic soda and sodium carbonate use for removing of hardness of water.
- CFB stands for Corrugated Fiber Board and the cartons come in 3 ply or 5 ply.
- Chantancy is excellent variety of carrot for canning.
- Chekurmanis is commonly called as "Multi Vitamin Greens"
- Chromatography was introduced by Tswal.
- Clinching is related to canning.
- *Clostridium* and *Bacillus* are spore forming bacteria.
- CO_2 Absorber- Lime, active charcoal, magnesium oxide.
- Codex Alimentarious Commission- body responsible for development of international standards for food safety and consumer protection.
- Concentration of CO_2 above 1% delay ripening in vegetables.
- Conduction is the heat transfer that takes place from one particle to the next particle without the movement of the particles in food e.g., heating/drying of solid food.

- Controlled Atmosphere Storage was invented by Kidd and Wert.
- Convection is the heat transfer that takes places with the movement of particles themselves.
- Conversion of unsaturated fat into saturated fat is by the process known as Hydrogenation.
- Cucumber pickle ranks first in world.
- Dark green leaves of leafy vegetable contain small amount of "Omega-3" fat.
- Decortication is the method related with breaking of pits of stone fruits.
- Development of "Off flavour" in fat is known as rancidity.
- Drink of date palm is known as "Dibbis".
- During raisin preparation, processed grapes are dried till water reduces to 13-15%.
- Ethylene absorbents e.g., purafil extend storage life of tomatoes by absorbing the ethylene evolved in the package during storage.
- Ethylene Absorber-Potassium Permaganate, Allumina pellets oxidizes, vermiculite, silica gel, hydro carbon, silicones (Phenyl-methyl silicone).
- Excellent variety of carrot for canning is Chantenay.
- Extrusion is a size enlargement process in which pressures of 1500-70,000 KPa and temperature of 52-199°C are maintained.
- Facultative anaerobes are basically aerobic but can grow occasionally under anaerobic conditions.
- FAO had categorized food processing into 3 sector i.e., primary processing, secondary processing and tertiary processing.
- Fat soluble components and essential fatty acid are lost during irradiation and such products (e.g., dairy product) are therefore instable to irradiate due to development of rancid off flavour which is also accelerated by the presence of oxygen, therefore some of the meat product are irradiated under vacuum.
- Fibril theory of jelly formation is given by W.V. Cruess.

- Filtragol and Maltolysin enzyme required for clarification of unfermented apple juice.
- Firmness (a degree of softness or crispness) of fruits is measured by Penetrometer.
- Flavonoides also were known as a vitamin P (for permeability because they were shown to have effect on the permeability of blood capillaries but were not dietary essentials).
- Flavour is a combination of taste and aroma.
- Flower like rose, carnation, chrysanthemum etc. are highly ethylene sensitive while flowers like gladiolus, tuberose, etc are low to insensitivity to ethylene.
- Food irradiation is also cooled as cold sterilization.
- Food poisoning is caused by *Clostridum botulinum* and *Staphylococcus aureus*.
- For batter MAP uses Polypropylene (PP) packaging film.
- For better quality of onion and garlic, irrigation should be stopped 3 weeks before harvesting.
- For dehydration, onion should be white, round, without any green patches, TSS 15-20°Brix, low ratio of reducing to non-reducing sugar, good keeping quality and 80% moisture.
- For each 300 feet increase in altitude, the boiling point of water decreases about 1°F.
- For enhancing shelf life of apples it is recommended to dip in 4.0% calcium chloride.
- For low sugar content, potato tubers are stored at 10°C.
- Freezing point depression of a solution is mastered with Calorimeter.
- Freezing preserve the food through inactivation of microbial principle.
- Fruit and vegetable processing for the first time was started in a systematic manner in the year 1857.
- Fruit drinks are known as imitation foods.
- Fruit proportion in preserve should be 55%.
- Fruits and vegetables with yellow coloured skin are rich in Vitamin-A.

- Fruits in which colouring matter is insoluble in water are packed in plain cans.
- Glucose oxides convert glucose into gluconic acid while Glucose isomerase converts glucose into fructose.
- GM (Genetically Modified) food are known to contain new proteins produce by the recombination DNA technique and also called as transgenic or biotech food.
- GM may be toxic and allergent.
- Gulkand is prepared by mixing rose petals and sugar in the ratio of 1:1.
- HACCP concept was developed in the year 1960 to ensure safety of food products.
- Halophiles are salt tolerant and osmophiles are sugar loving.
- Hardness of water is due to presence of dissolved calcium and magnesium salt.
- Harvesting of lotus rhizomes is done in the month of October.
- Heat of respiration is an energy released by the produce as it respires during storage. The warmer the produce, the more heat of respiration generated.
- Heat peeled fruits absorbs sugar more readily than fruit peeled by other methods.
- *Helichrysum* flower is suitable for air drying.
- High dose (>10 KGy) are employed for sterilization of packaging material and packed meat, poultry products to make them self stable without refrigeration (25-70 KGy) and sterilization of hospital diets (25-70 KGy).
- Highest number of cold storages (2012) out of the total in the country are being used for potato in India.
- Homo-fermentative bacteria produce only lactic acid while hetro-fermentative bacteria produce lactic acid, acetic acid, CO_2 and ethanol.
- Horticulture contributes 28.5% of GDP in agriculture and 52% of export share in agriculture (2010-11).
- Hypobaric storage concept is given by Burg and Burg in 1966.
- Hypobaric storage is also known as Low pressure storage.

- Hypobaric storage is mostly used for cut flower storage.
- If the food is heated the process is called as extrusion cooking.
- In canned vegetable products, *Colstridium botulinum* is common infection.
- In citrus fruits which morphological part is known "Albedo" to be rich in pectic substances and hemicelluloses.
- In climacteric fruits ethylene biosynthesis can be reduced by the application of Aminoethoxyvinyl glycine.
- In fruit canning and pickling salt act as antioxidant.
- In fruit juices, losses of vitamin-C and carotene are minimized by de-aeration.
- In garlic and onion, a bioflavonoides i.e., Quercetin is present which provide protection against cancer and heart diseases.
- In hypobaric storage fruit are stored at low pressure (0.2-0.5 atm) & low temperature (59-75°F or 15-23.9°C).
- In jelly making, pectin acts as a stabilizer.
- In jelly, red coloure formation takes place due to charring of sugar.
- In oven drying most of flowers are dried for 48-72 hours.
- In pickle white cottony surface developed due to *Rhyzopus* bacteria.
- In the papaya found maximum post harvest losses.
- India is the fruit and vegetable basket of the world.
- India produces 54% of world's mango, 23% banana, 24% cashew nuts, 36% green peas and 10% onion production.
- Indicator bacteria on freezing is *Enterococci*.
- Individual Quick Freezing (IQF) is commonly practiced in peas.
- "Khalpiis" prepared by fermentation of cucumber.
- Kidd (1934) attributed the sudden rise in respiration of some fruits during ripening to the presence of activated fructose in the cytoplasm.
- Lime pickle is the national pickle of India.
- Lipase enzyme responsible for hydrolysis of fat.

- Liquid oozes out during thawing are called leakage.
- Litchi fruits are harvested in bunches with few leaves to avoid rupturing.
- Louis Pasteur proved that micro organisms are the real cause of spoilage.
- Low dose (<1KGy) are employed for disinfestations of grain and fruits for insect and larva.
- LTLT pasteurization is also known as "Batch Pasteurization".
- Lycopene oxidizer and turns brown when comes in contact with iron. It also forms black compounds with the tannins present in tomatoes or the spices which are used for product preparation.
- Magnetic field intensity between 5-50 tesla, with short pulses of 25μ sec up to few minutes and frequency of 5-500 KHz are required for inactivation of microorganism.
- Magnetron is device to convert electric energy at low frequencies into an electromagnetic field with centers of positive and negative charge that change direction billions of times per second. This causes the instantaneous penetration and heating of food.
- Maleic Hydrazide (MH) inhibits sprouting in turnips, radish, onion, potatoes and carrot at a concentration of 2500ppm, when applied 10-14 days before harvesting.
- Malnutrition is condition caused by under and over nutrition.
- Marsh spot in peas is due to deficiency of Magnese (Mn).
- Methionine is the precursor in ethylene biosynthesis.
- Microorganisms die due to irreversible heat denaturation of enzyme, protein, nucleic acid and other cellular constituents vital to cell metabolism and reproduction result in cell death.
- Mild swell at one or both end of a can is called springer.
- Ministry of food processing was established in the year 1996.
- Mixing of two or more different products to get desired final products to get desired final product is called blending.
- Mohr titration method is used for estimation of salt.

- Most suitable packaging material for cut flowers is card board boxes.
- Mushrooms are very rich in lysine and trytophane, the two essential amino acids deficient in cereals.
- Mushrooms contain 20-35% protein (dry weight basis) which is higher than vegetables and fruits and is so high quality.
- Mycobacterium bacteria used to absorb ethylene from storage chamber.
- Mycotoxins are fungal metabolites highly toxic to many animals and potentially toxic to human beings.
- Niacin is the only vitamin which is stable in all the conditions including heat.
- No. 1 processing vegetable tomato is treated as protective food.
- Non-enzymatic browning is caused by reactions between nitrogenous compounds and sugar; nitrogenous compounds and organic acids; sugars and organic acids; and among organic acids themselves, these reaction are known as Maillard reactions.
- Washing soda is Sodium Carbonate.
- Okra is the important source of iodine to human diet.
- One kg essential oil is obtained from 3-4 tones rose petals.
- One ton of refrigeration is defined as the energy removed from one ton (2,000 lbs) of water so that it freezes within 24 hours at 0°C (32°F). It is equivalent to 2,88,000 Btu/day or 12,000 Btu/h.
- One tonne of refrigeration is defined as the energy removed from one metric tonne (1000kg) of water so that it freezes within 24 hours at 0°C.
- Only y- rays from cobalt-60 or caesium -137, x-ray generated by a machine at maximum energy of electronics generated by a machine at a maximum energy of 10 gama volts can be used which are to low to induce radio activity in any material including food, exposed to them.
- Orange and pineapple nectar should be 40% minimum fruit juice container.
- Organisms insensitive to freeze are *Bacillus* and *Clostridium*.

- Oscillating magnetic field couples energy in to the magnetically active part of large critical molecules such as DNA, resulting in to break down of covalent bonds in the DNA molecules and inhibition of the growth of microorganism
- Oxalic acid is present in Carambola.
- Oxygen Absorber-Iron Powder, FeO, Fe_2O_3, Fe_3O_4 and Iron hydroxide.
- Pectinol enzyme used for clarifying of juice.
- Pepsin is present in papaya.
- Peroxidase used as an index of adequacy of blanching.
- Phytochemicals are chemicals that are produced by plant butr not consider essential nutrient (protein, carbohydrate, fats, minerals and vitamins)
- Picowaved: this term is used to label the food which has been treated by low level of ionizing radiation.
- Pink colour onions are used for dehydration.
- Polyhydroxy butyrate is biodegradable plastic.
- Post harvest dip in 2000 ppm MH reduces senescence of lettuce, retard degredation of chlorophyll in beans and increase the shelf life, inhibits the synthesis of solaninein potatoes under storage.
- PPO complex is used as an indicator for blanching.
- Protein denaturation leads to increased toughness of the frozen products.
- Psychrometric is the study of inter relationship between temperature and humidity of the air.
- Pycnometer is an instrument used to measure specific gravity.
- Radappertization: It is equivalent to radiation sterilization/ commercial sterility. It is the destruction of practically all microorganisms achieved by high dose of irradiation
- Radiation is the heat transfer that takes places by electromagnetic waves.
- Radiowaves have 30 m to 30 km wave length
- Refrigerated Storage- Its one of the first successful cold storage for apple was built in 1856 by Reverend Benjamin Nycein.

- Relative humidity is a ratio of partial pressure of water vapour in air to pressure of saturated water vapour at same temperature × 100.
- Respiration rate is the rate of consumption of O_2 or evolution of CO_2 from a commodity. Unit=ml CO_2 $Kg^{-1}h^{-1}$.
- Retortable pouches are made of three layers i.e. polyester aluminum foil and polypropylene.
- Rhubarb is acidic in nature.
- Ripening inhibitor (rin) gene is associated with tomato.
- RQ is expressed as amount of CO_2 evolved to amount of O_2 absorbed.
- Saccharine is the first synthetic sweetener which is 300 times sweeter than sucrose.
- Saddington describes canning methods for 1st time in1807.
- Salt concentration in pickle is measured by solometer.
- Salting a method of preservation is chemical in nature.
- Senescence in leafy vegetables can be retarded by the application of Benzyle Adenine.
- Separate Ministry of Food Processing was established in the year 1999.
- Silver ions are effective antagonists of ethylene action on fruit because it is reacts with ethylene to form an ineffective complex.
- "Sinki" a fermented product is prepared from radish tape root.
- Sodium Benzoate acts between 2.5-4.0 pH ranges.
- Sodium bicarbonate is Baking soda.
- Softening of fruit during ripening is due to the breakdown of pectic substances.
- Sorbates are salt of sorbic acid and are safer than other preservatives.
- Spoilage of canned product caused by bacteria which produce acid but no gas is called Flat sour.
- Starch changes in glucose in guard cell at pH-7.
- Strawberry fruit have 98% edible portion.

- Sugar acts as preservative above 66% while salt above 12% concentration.
- Sugar concentration can be measured by hydrometer and refractometer.
- Sugarcane and sugar beet are source of sugars.
- Sulphur resistant lacquers are golden coloured and can coated with it are called C enamel cans and are suitable for packaging non-acid products like peas, corn, lima beans, red kidney beans, etc.
- Sweating is the process of stacking dried fruit in boxes/bins to equalize the moisture contents before final packaging.
- *Tagetes minuta* species of marigold, use for extraction of essential oil.
- Tannin is responsible for astringancy in Brinjal and Bottlegourd.
- Terrarium is transparent race for keeping flowers.
- The "World Food Day" is celebrated every year on 16, October.
- The 1st commercial product processed by HPP was offered in Japan in the year 1920.
- The anti-fungal factor in onion is phenolic compound known as Catechol.
- The coloured portion of peel in citrus is known as Flavedo.
- The combination of osmotic and vacume drying is referred to osmovac process.
- The concentration of oxygen at which the anaerobic respiration starts is known as Extinction Point.
- The efficiency of conversion of sugar to alcohol is about 90%.
- The energy liberated during respiration of stored fruit is known as "Vital heat".
- The first commercially grown genetically modified hole food crops was tomato (called as **falvr savr**) which was made more resistant to rooting by Californian company calgene. Variant of the flavr savr was to produce tomato paste.
- The Horticulture Society of India was established in the year 1942.

- The inner side of the can is coated with lacquer which prevents discoloration of product and this process is known as lacquering.
- The juice of pomegranate fruit is valued for treatment of leprosy disease.
- The lactic acid bacteria are salt tolerant and they can flourish in a brine of about 8-10% salt concentration.
- The lethal dose of HCN for human being is estimated to be 0.5-3.5 mg per kg body weight.
- The light treatment can be used for inactivation of microorganism at the surface of food and packaging materials and with transparent bulk liquids, solids and gases.
- The limiting amino acid in green vegetables is methionine.
- The main mechanism of microorganism are killed are (i) effect of DNA and RNA molecule of microorganism, the DNA double helix fails to unwind and microorganism cannot reproduce (ii) change the structure of cell membrane, (iii) effect the metabolic enzyme activity.
- The maximum permissible limit for the use of anti-caking agents is 2%.
- The only inorganic acid widely used in the food industry is phosphoric acid.
- The oriental fermented food 'Poi' is made from taro corms.
- The overheating of canned foods by remaining too hot while being stacked in the warehouse is called stack burning.
- The process of "wet steam formation" in the boilers is called priming.
- The process to convert unsaturated fats into saturated fats with addition of hydrogen is called hydrogenation.
- The process used to hold dehydrated fruits in bins to equalize moisture are called sweating.
- The process of adjusting the sugar content for wine making is referred as amelioration.
- The product prepared by blending ginger juice or its essence with water and nutritive sweeteners is known as "ginger bear".

- The quickest method of peeling is Flame peeling.
- The red pigmentation of apple is due to cyanidin-3-galactoside.
- The rules and produce for handling GM food in India are being monitored by IBSC (Institutional Bio-Safety Committee) and Review Committee on Genetic Manipulation (RCGM).
- The starch-iodine test for judging apple maturity results in the formation of dark bluish black colour.
- The term 'climacteric' is first used by Kidd and West (1927).
- The threshold level of ethylene is 0.03 μL/L in fruits and vegetables.
- The time required to destroy 90% of micro-organism in preserved products is D value.
- The word market has been derived from Latin word "Marcatus" meaning merchandise.
- The word protein is derived from "Greek".
- Thermal death time is the amount of time required to kill a specific number of microbes at a specific temperature.
- Thermo-radiation: This is the simultaneous or sequential combination treatment of heat and radiation, use full for food prevention.
- Those properties of food stuffs apprehended by the eyes and by the skin muscle senses of the mouth, including the roughness, smoothness graininess is known as texture.
- Tomato is the first commercially grown genetically modified food crops.
- Tooty fruity is a process product prepared from bottle gourd.
- A total sugar content of impregnated fruit is kept at about 75% to prevent fermentation.
- Trans-membrane potential (TMP) is the difference in potential between bacteria cell and its external environment. TMP of cell= 1 volt.
- United Kingdom is the largest importer of dry flowers from India.
- Usually a low frequency alternating heating current (50-60 Hz.) is used in ohmic heating (electric conducting liquid particulates the size up to 25 mm).

- Vacuum pre-cooling is more suitable for leafy vegetables.
- Van't Hoff, a Dutch chemist described that the rate of a chemical reaction almost doubles for every 10°C rise in temperature and gave a temperature quotient Q_{10}.
- Vital Heat is the energy is released in the form of heat during respiration.
- Vitamin-B_1 is an appetite vitamin.
- Vitamin-C mostly destroyed during freezing due to oxidation.
- Water constitute of protoplasm helps in transit of food.
- Water percentage in fresh rhizome of lotus is 80.0%.
- White cabbage is more suitable for making sauerkraut than winter cabbage because white cabbage has low content of polyphenols.
- Zero energy cool chambers developed by S.K. Roy and Khurdiya in 1982.
- Zero energy cool chambers maintain a temperature 15-17°C less than outside temperature.

28

Multiple Choice Questions

1. Carbohydrate content in potato is :

   (A) 12% (B) 22%

   (C) 32% (D) 42%

2. Ingredient gelling agent, emulsifiers and stabilizers in vegetable product processing :

   (A) Gum (B) Sodium benzoate

   (C) Pectin (D) A and B

3. Micro organism bacteria, yeasts and mould are responsible in processing :

   (A) Improve the storage life

   (B) Help in preservation

   (C) Spoil the produce

   (D) Improve the quality

4. Which of following is less sensitive to ethylene?

   (A) Gerbera (B) Carnation

   (C) Orchid (D) Rose

5. Metabisulphite used as preservative of finished product :

   (A) 0.7 gm/kg (B) 0.5 gm/kg

   (C) 0.6 gm/kg (D) 0.3 gm/kg

6. Freeze drying also called :
   (A) Lypholisation (B) Sublimation
   (C) None of this (D) Both A and B

7. Which refrigerant is commonly used is used in cold storage in our country?
   (A) Ethylene (B) Carbide
   (C) Ammonia (D) Sodium Bcnzoatc

8. The destruction of all pathogens in food by thermal processing is known as :
   (A) Brining (B) Pasteurization
   (C) Sterilization (D) Drying

9. Two conflicting processes of growth-promotion and retardation of metabolism are found in :
   (A) Fruit (B) Vegetable
   (C) Flower (D) Foliage

10. Potatoes suffer from internal break down known as 'black heart' when stored at the temperature :
   (A) 0^oC (B) 5^oC
   (C) 2^oC (D) 4^oC

11. Lye solution contains caustic soda :
   (A) 5% (B) 3%
   (C) 1-2 % (D) 4%

12. Excessive shrinkage and early sprouting occur if potatoes are held at temperature above :
   (A) 8^0C (B) 10^0C
   (C) 5^0C (D) 12^0C

13. Mango variety having strong flavour is :
   (A) Dashari (B) Sindhu
   (C) Langra (D) Fazli

14. A concentrated tomato juice or pulp without skin and seed and containing not less than 25 per cent of tomato solids, is known as :

(A) Tomato paste (B) Tomato puree
(C) Tomato ketchup (D) All of this

15. The process of heat treatment for food products at which all the micro organisms are destroyed, is called :

(A) Brining (B) Pasteurization
(C) Sterilization (D) Drying

16. Process of preservation of food in common salt or vinegar is called :

(A) Syruping (B) Pickling
(C) Both of this (D) None of this

17. The process of removal of field heat in horticultural produce is called as :

(A) Evaporation (B) Transpiration
(C) Precooling (D) Evapotranspiration

18. Pineapple variety suitable for canning is :

(A) Queen (B) Kew
(C) Mauritius (D) Cayenne

19. Preserving of food by heat is known :

(A) Drying (B) Freezing
(C) Dehydration (D) Processing

20. Strength of the brine solution is measured by __________.

(A) Salometer (B) Penitometer
(C) Manometer (D) Refractometer

21. Water activity in product is measured by __________.

(A) Evaporometer (B) Penitometer
(C) Hygrometer (D) Jellmeter

22. Sorbic acid is considered ______preservative due to physiological point of view.
    (A) Harmful (B) Safe
    (C) Moderate (D) None
23. Richest source of Riboflavin is :
    (A) Papaya (B) Mango
    (C) Bael (D) Karonda
24. Fruit crop exported on large scale from India is : (JRF 1997)
    (A) Banana (B) Mango
    (C) Citrus (D) Papaya
25. Maximum vitamin A content : (JRF-2010)
    (A) Carrot (B) Drumstick leaves
    (C) Pumpkin (D) Pea
26. The fruit having richest source of riboflavin is :
    (A) Banana (B) Bael
    (C) Mango (D) Anola
27. Which one of the following is the permanent method of preservation of fruits :
    (A) Canning (B) Moisture exclusion
    (C) Sterlization (D) Pasteurization
28. For sterilization, the temperature should be :
    (A) 80^0C (B) 74^0C
    (C) 65^0C (D) Above 100^0C
29. Father of canning is :
    (A) Nicholas Appert (B) Needham
    (C) Spallanzani (D) Peter Durand
30. Refractometer is used to determine :
    (A) Acid (B) Texture
    (C) TSS (D) Colour

31. Richest source of iron is :

(A) Mango (B) Bael

(C) Pomegranate (D) Dry Karonda

32. Vapoure heat treatment is done in : (JRF-2010)

(A) Mango (B) Apple

(C) Pear (D) Pineapple

33. Which one of the following spice crops is largely exported from India : (JRF-1997)

(A) Cardamom (B) Black pepper

(C) Coriander (D) Turmeric

34. Physiological process responsible for rapid deterioration of freshly harvested fruits and vegetables is : (JRF-2005)

(A) Photosynthesis (B) Evaporation

(C) Transpiration (D) Respiration

35. Guava is a rich source of : (JRF-1997)

(A) Vitamin-C (B) Vitamin-A

(C) Iron (D) Protein

36. Which one of the following is a Climacteric type of fruit :

(A) Banana (B) Citrus

(C) Litchi (D) Grape

37. Which of these can kill micro-organisms in food : (JRF-2005, 2012)

(A) Sugar (B) KMS

(C) Heat processing (D) Citric acid

38. Pungency of onion is due to the presence of : (JRF-1999, 2004, 2005)

(A) Allylpropyl disulphide (B) Solanin

(C) Isothiocynate (D) ——

39. Fruit of Rose is known as : (JRF-2005)

(A) Hip (B) Drupe

(C) Berry (D) Blastula

40. Potato is a good source of : (JRF-1999)
    (A) Pectin (B) Starch
    (C) Carotene (D) Tannin
41. Edible part of pomegranate : (JRF-2010)
    (A) Mesocarp (B) Endocarp
    (C) Epicarp (D) Aril
42. 'Champain' is a type of : (JRF-1997)
    (A) Pickle
    (B) Unfermented fruit beverages
    (C) Vinegar
    (D) Wine
43. "Most suitable packaging material" for cut flowers is :
    (A) Wooden boxes (B) Plastic boxes
    (C) Cardboard boxes (D) Caretes
44. In normal air how much ethylene is present :
    (A) 1ppm (B) 100ppb
    (C) 3-5 ppb (D) 5-10ppb
45. How much ethylene concentration is detrimental to cut flowers?
    (A) 10-20 ppb (B) 30-50 ppb
    (C) 50-100 ppb (D) 1 ppm
46. Which one of the following is a non-climacteric fruit? (JRF-1997)
    (A) Musk melon (B) Cucumber
    (C) Water melon (D) None of the above
47. What is the minimum content of juice in RTS beverages : (JRF-2005)
    (A) 10% (B) 15%
    (C) 20% (D) 25%
48. Guava is a good source of : (JRF-1999)
    (A) Pectin (B) Starch
    (C) Carotene (D) Tannin

49. Emission of Ethylene during transportation of cut flowers cause a disorder which called as :

(A) Bud opening (B) Sleepiness

(C) Bent neck (D) Calyx splitting

50. Tomato sauce must have TSS : (JRF-2005)

(A) 28% (B) 68%

(C) 25% (D) 30%

51. Best storage temperature for gladiolus spike is : (JRF-1997)

(A) 2-4°C (B) 4-6°C

(C) -1°C (D) 13°C

52. Guava jelly is preserved by : (JRF-1999)

(A) KMS (B) Citric Acid

(C) Sodium benzoate (D) Pectin

53. In tomato fruit processing, pH of the fruit should be : (JRF-2006)

(A) > 4.5 (B) < 4.5

(C) < 3.5 (D) >7

54. First commodity for which grading and marketing rules were framed is :

(A) Tomato (B) Mango

(C) Grape (D) Onion

55. Papaya contains........... unit of Vitamin-A. (JRF-1997)

(A) 2010 IU (B) 2020 IU

(C) 4800 IU (D) 4200 IU

56. Vitamin-C is maximum in : (JRF-1997)

(A) Aonla (B) Orange

(C) Guava (D) Mango

57. Yeast is used in making : (JRF-1999)

(A) Cider (B) Cordial

(C) Nectar (D) RTS

58. What is the permissible limit of SO_2 in raisin grapes : (JRF-2005)
    (A) 500 ppm (B) 1000 ppm
    (C) 250 ppm (D) 750 ppm
59. Cauliflower curds can be stored for a month at :
    (A) O°C with 85-90% RH
    (B) 15°C with 60-80% RH
    (C) 15 °C with 60-65% RH
    (D) 20 °C with 50-70% RH
60. Pectin test is done by : (JRF-1997)
    (A) Acid (B) Alcohol
    (C) Both A and B (D) Jelmeter
61. Bacteria is responsible for the production of : (JRF-1999)
    (A) Nectar (B) Cider
    (C) Vinegar (D) Wine
62. Oleoresin is extracted from : (JRF-2005, 2007)
    (A) Garlic (B) Capsicum
    (C) Onion (D) Cassava
63. For curing, sweet potato are kept for 10 days at :
    (A) 25 °C and 85% RH (B) 40°C and 70% RH
    (C) 80 "C and 30% RH (D) 30 °C and 80% RH
64. Central Coffee Research Institute is located at : (JRF-1997)
    (A) Bangalore (B) Delhi
    (C) Balanhur (D) Muzeferpur
65. Which of the following amino acid is the precursor of ethylene production? (JRF-1999)
    (A) Methionine (B) Valine
    (C) Cystine (D) Arginine
66. "Elephant's Foot Yam" is rich source of Vitamin :
    (A) A and B (B) B and C
    (C) C and D (D) Only B

67. FPO permits the use of only two chemical as preservative : (JRF-2005)
    - (A) Boric acid and Sodium chloride
    - (B) Potassium meta bisulphide and sodium benzoate
    - (C) Acetic acid and citric acid
    - (D) Potassium disulphide and Sodium carbonate

68. Aroma in coffee is due to : (JRF-1997)
    - (A) Drying
    - (B) Roasting
    - (C) Fermentation
    - (D) All of these

69. Standard unit used in expressing refrigeration capacity is : (JRF-1999)
    - (A) Kg
    - (B) Ton
    - (C) Jule
    - (D) Ampere

70. Tomato fruits for canning are harvested at :
    - (A) Mature green stage
    - (B) Red ripe stage
    - (C) Immature green stage
    - (D) Half-ripe/pink stage

71. Chilli are good source of : (JRF-1997)
    - (A) Carbohydrate
    - (B) Iron
    - (C) Vitamin-C
    - (D) Calcium

72. The yellow pigment in papaya is : (JRF-1999)
    - (A) Carotene
    - (B) Caricacin
    - (C) Xanthophylls
    - (D) Caricaxanthin

73. For making a fruit jelly, the correct proportion of pectin and sugar should be : (JRF-2005)
    - (A) Pectin 0.8-1%, sugar 65-68%
    - (B) Pectin 2-3%, sugar 50-55%
    - (C) Pectin 5-6%, sugar 70-75%
    - (D) Pectin 0.2-0.3%, sugar 55-60%

74. Pink colour of onion is due to : (JRF-1997)

(A) Qucertine (B) Anthocyanins

(C) Carotene (D) Xanthophylls

75. In which of the following crops curing is done for ripening : (JRF-2008)

(A) Guava (B) Apple

(C) Orange (D) Banana

76. A treatment after harvesting by using water to restore turgidity is known as :

(A) Pulsing (B) Hardening

(C) Loading (D) Pre-cooling

77. Treatment of cut flowers after harvesting with high concentration of sugar is known as :

(A) Pulsing (B) Hardening

(C) Loading (D) Pre-cooling

78. For pulsing optimum room temperature is required :

(A) 10-15°C (B) 15-20°C

(C) 20-25°C (D) 25-30°C

79. Optimum pH of holding solution should be :

(A) 2-3 (B) 4-5

(C) 5-6 (D) 6-7

80. Commonly used for prolonging vase-life of cut flower is : (JRF-199)

(A) Potassium Nitrate (B) Sucrose

(C) B-Nine (D) MH

81. The bitterness is colocasia corms is felt due to the presence of : (JRF-2005)

(A) Capscicin

(B) Calcium carbonate

(C) Sodium chloride

(D) Calcium oxalate

82. Melons for distant marketing arc picked at :

(A) Half-slip stage (B) Full-slip stage

(C) Green mature stage (D) None of these

83. Which type of can is used for packing of spinach, cauliflower and sweet potato : (JRF-1997)

(A) AR cans (B) SR cans

(C) Plain (D) All of the above

84. Which one is belong to non-climacteric group : (JRF-1999, 2005)

(A) Pineapple (B) Avocado

(C) Persimmon (D) Mango

85. For distant marketing, tomato fruits are harvested at :

(A) Immature green stage (B) Mature green stage

(C) Turning stage (D) Red ripe stage

86. Which of the following commercial formulation are used as preservative :

(A) Seven Up (B) Petal life

(C) Oasis (D) All of these

87. Best maturity indice of grape is : (JRF-1997)

(A) Acidity % (B) Colour change

(C) Specific gravity (D) TSS

88. The active principle of clove oil is : (JRF-2008)

(A) Eugenol (B) Aldehyde

(C) Macine (D) Oleum

89. In which group will classify apple according to its respiration rate? (JRF-1999)

(A) Low (B) Moderate

(C) High (D) Very high

90. CO_2 and ethylene are removed when atmospheric pressure is :

(A) 0.1 atm (B) 1 atm

(C) 10 atm (D) 100 atm

91. The major chemical compounds responsible for the specific aromas of most ripening fruits are : (JRF-2005)

(A) β-carotene and its derivatives

(B) Esters of aliphatic alcohol's and short chain fatty acids

(C) Isoamyl acetate

(D) Ascorbic acid

92. For low sugar content, potato tubers are stored at :

(A) 5°C (B) 10 °C

(C) 15°C (D) 20°C

93. Which one of the following is ripening hormone : (JRF-1997, 2009)

(A) Ethylene (B) IAA

(C) GA_3 (D) ABA

94. Which of the following is the major flower grown for fragrance in India? (JRF-2007)

(A) Tuberose (B) Jasmine

(C) Gladiolus (D) Chrysanthemum

95. Bent neck in rose is very common when grow at :

(A) Low light intensity

(B) Medium light intensity

(C) High light intensity

(D) None of these

96. In which group will you classify mango according to its ethylene production rate? (JRF-1999)

(A) Low (B) Moderate

(C) High (D) Very high

97. Metallic coloure are : (JRF-2005)

(A) Favoured in preservation

(B) Strongly recommended in preservation

(C) Banned in preservation

(D) Recommended in preservation

98. For long-term storage, potato should be stored at :

(A) 0-5°C (B) 5-10°C

(C) 10-15°C (D) 15-20°C

99. Minimum TSS required for Jam is : (JRF-1997, 2010)

(A) 55% (B) 65%

(C) 68% (D) 72%

100. Leafy vegetables are usually rich source of : (JRF-2008, 2010)

(A) Iron (B) Protein

(C) Carbohydrate (D) Fat

101. Ideal cool temperature for banana storage is : (JRF-1999)

(A) 1-3^{0}C (B) 5-6^{0}C

(C) 7-10^{0}C (D) 13-14^{0}C

102. Which of following block xylem of cut flower?

(A) Bacteria (B) Mould

(C) Yeast (D) All of these

103. Maximum "Vit-C" content is found in : (JRF-2005)

(A) Barbedos cherry (B) Aonla

(C) Guava (D) Citrus

104. Which of following is effective bactericide?

(A) Silver Nitrate (B) HQC

(C) Cobalt Nitrate (D) Sugar

105. Tomato fruits for processing, are picked at :

(A) Pink stage (B) Hard ripe stage

(C) Over ripe stage (D) Mature stage

106. The main preservative used to preserve tomato ketchup is : (JRF-2008, 2009)

(A) Propionic acid (B) KMS

(C) Sodium Benzoate (D) Niacin

107. How much juice is required for making fruit jelly? (JRF-1997)

(A) 20% (B) 30%

(C) 45% (D) 40%

108. Organization related to Agricultural Marketing is : (JRF-2010)

(A) NABARD (B) UNESCO

(C) FCI (D) NAFED

109. Ideal storage temperature of cucumber is : (JRF-1999)

(A) 0-3^{0}C (B) 3-4^{0}C

(C) 10-13^{0}C (D) 13-15^{0}C

110. What are the approximate post harvest losses of fruits in India? (JRF-2009)

(A) 20% (B) 30%

(C) 40% (D) 50%

111. Which of the following is most important ethylene inhibiter :

(A) Silver Thiosulphate (B) Sugar

(C) Citric acid (D) Chloride

112. Polyphenol oxydase enzyme inactivate at the temperature of : (JRF-2007)

(A) 80-90^{0}C (B) 90-100^{0}C

(C) Above 100^{0}C (D) None of them

113. Major alkaloid present in cymbopogon : (JRF-2006)

(A) Quinine (B) Citronellal

(C) Vincristine (D) None of the above

114. Kalipak is prepared from : (JRF-1997)

(A) Mature Green nuts of areca palm

(B) Immature Green nuts of areca palm

(C) Over ripe Green nuts of areca palm

(D) All of these

115. Which is most common germicide for flower preservation :
(A) Silver Nitrate (B) Salt of HQ
(C) Aluminum Sulphate (D) Silver Thiosulphate

116. The pigment present in Jamun is : (JRF-1999)
(A) Anthocyanin (B) Carotenoid
(C) Lycopens (D) Xanthophyll

117. Ripening of mango commercially done by : (JRF-2010)
(A) KNO_3 (B) Ethrel
(C) Calcium carbide (D) By Smoking

118. The Limiting Amino acid in green vegetables is :
(A) Arginine (B) Lysine
(C) Methionine (D) Tryptophan

119. The sparkling clear liquid free from all suspended solid is known as: (JRF-1997)
(A) Ready to serve (B) Squash
(C) Syrup (D) Cordial

120. Palm oil is obtained from which part of the fruit? (JRF-2008)
(A) Epicarp (B) Mesocarp
(C) Pericarp (D) Endocarp

121. The gas released during fruit ripening is : (JRF-1999)
(A) Methane (B) Ethylene
(C) Acetylene (D) Oxygen

122. Curing done in which crop? (JRF-2010)
(A) Potato (B) Tomato
(C) Onion (D) Brinjal

123. Which of the following is a ethylene absorbent? (JRF-2006)
(A) $KMnO_4$ (B) KNO_3
(C) K_2SO_4 (D) KCl

124. Which is the staple vegetable in Indian diet :

(A) Tomato (B) Cauliflower

(C) Potato (D) Chilli

125. Which is not ethylene inhibiter :

(A) AVG (B) MVG

(C) AOA (D) NAA

126. Treatment of fruit and vegetables by slight temperature for small time is known as : (JRF-1997, 2005)

(A) Pasteurization (B) Blanching

(C) Sterilization (D) Cooking

127. Which of following strongest senescence stimulator :

(A) ABA (B) BA

(C) IAA (D) PBA

128. Which of the following property of pectin is used in commercial preparation of jam and jellies? (JRF-2008)

(A) Causes cloud formation

(B) Decreases drastically during ripening

(C) Can form gel with sugar and acid

(D) Can be degraded with depolymerizing enzymes

129. Which one of the following statements is correct with respect to nutritive value? (JRF-1999)

(A) Plenty of water should be added while cooking vegetables

(B) Vegetable should be cooked in strong heat

(C) Vegetable should be peeled thoroughly before cooking

(D) Vegetable once cooked should not be heated

130. Spinach is rich in :

(A) Vitamin A (B) Vitamin B

(C) Vitamin C (D) Vitamin E

131. Which of following use as a wetting agent in flowers and foliage :

(A) Tween-20 (B) Soap

(C) Washing powder (D) All of these

132. Which one of the following methods is the permanent method of preservation? (JRF-1999)

(A) Sterilization

(B) Dehydration

(C) Pasteurization

(D) All of the above

133. Impregnation of cut flowers is very common in :

(A) Aster (B) Gerbera

(C) Carnation (D) All of these

134. Exudation of fluids in jams and jellies is due to the phenomenon known as : (JRF-2008)

(A) Synersis (B) Osmosis

(C) Retrogradation (D) Permentation

135. Which of following pot plant is highly sensitive to ethylene?

(A) Euphorbia (B) Scindapsus

(C) Cordyline (D) Nephrolepis

136. Which of the following vegetable is known as 'God sent vegetable"? (JRF-2008)

(A) Pea (B) Brinjal

(C) Lima bean (D) Winged bean

137. Coloured product preserved by : (JRF-2010)

(A) KMS (B) Benzoic acid

(C) SO_2 (D) None of the above

138. Who is the largest importer of indoor plants?

(A) Netherland (B) Italy

(C) USA (D) Germany

139. Zero energy chambers are devices which are helpful to enhance shelf life of fruits because : (JRF-2007)

(A) Maintain low temperature as compared to ambient conditions

(B) Maintain high humidity as compared to ambient conditions

(C) Utilize no electricity

(D) All of these

140. Which bean is used for extraction of gum?

(A) Broad bean (B) Cluster bean

(C) French bean (D) Hyacinth bean

141. What pH is the dividing line between the acid and non-acid foods? (JRF-2008)

(A) 4.0 (B) 4.5

(C) 5.0 (D) 7.0

142. Highest quantity of carotene is available from : (JRF-1999)

(A) Colocasia leaf (B) Carrot

(C) Spinach leaf (D) Fenugreek leaf

143. Fermented beverage among the following is : (JRF-2010)

(A) Juice (B) Jelly

(C) Vinegar (D) Jam

144. Chillies are rich source of :

(A) Vitamin A (B) Vitamin C

(C) Vitamin A and C (D) Vitamin E

145. Daily requirement of "Vitamin-A" for an healthy adult is : (JRF-1999)

(A) 50-60mg (B) 4000-5000 IU

(C) 100 IU (D) 0.003-0.005 g

146. Red colour of carrot is due to :

(A) Lycopene (B) Anthocyanin

(C) Carotene (D) Quercetin

147. Which is the fermented product? (JRF-2009)

(A) Cider (B) Sauce

(C) Jam (D) Squash

148. The red colour of chilli fruit at ripening stage is due to the pigment : (JRF-2007)

(A) Carotene (B) Capsaicin

(C) Capsicutin (D) Capsanthin

149. A separate Ministry of Food Processing in the Government of India was created in the year : (JRF-1999)

(A) 1988 (B) 1950

(C) 1955 (D) 1968

150. Vegetables are subjected to drying after :

(A) Sulfuring (B) Sulphitation

(C) Blanching (D) None of these

151. Working principle of Pusa energy cool chamber :

(A) Evaporative cooling (B) Electricity supply

(C) Artifical cooling (D) None of the above

152. Non-climacteric fruits are those which : (JRF-1999)

(A) Will ripen properly only when harvested from the plant before ripening

(B) Will ripen properly only while attached to the plant

(C) Will show an abrupt increase in respiration rate at the ripening stage

(D) None of the above

153. Yellow coloured vegetables are rich source :

(A) Vitamin A (B) Vitamin B

(C) Vitamin C (D) Vitamin D

154. Vegetable are rich in : (JRF-2010)

(A) Protein (B) Carbohydrate

(C) Fibers (D) Vitamins

155. Cashew nut fat contains is : (JRF-1999)

(A) 53% (B) 63%

(C) 21% (D) 47%

156. Which is not moderately sensitive to ethylene?

(A) Dracaenes (B) Dieffenbachia

(C) Asparagus (D) Schefflera

157. Which is not highly sensitive to ethylene?

(A) Achimenes (B) Fuchsia

(C) Hibiscus (D) Primula

158. Ideal packaging is one which help in : (JRF-2007)

(A) Minimizing post-harvest losses

(B) Makes commodity attractive for consumers

(C) More fit for transportation

(D) All of the above

159. In tomato fruit processing, pH of the fruit should be :

(A) <4.5 (B) >4.5

(C) 4.5 (D) Neutral

160. Which is less sensitive to ethylene?

(A) Chrysanthemum (B) Kalanchoe

(C) Begonia (D) Beloperone

161. Cheapest method of preservation : (JRF-2010)

(A) Drying (B) Freezing

(C) Fermentation (D) Canning

162. Which of the following fruits is most suitable for canning? (JRF-2006)

(A) Mango (B) Pineapple

(C) Banana (D) Avocado

163. Green Glo is used for improving longevity of :

(A) Flowering Pot plant (B) Foliage pot plant

(C) Flowering plants (D) Fruits

164. Father of modern Refrigeration is :

(A) James Harrison (1851) (B) Gane (1934)

(C) Wade, N.L. (1984) (D) Wang (1986)

165. In vegetable blanching is done for : (JRF-2000)

(A) Removal of micro organism

(B) To inactivate the enzymes

(C) To retention of colour

(D) All of the above

166. Fruit for diabetes : (JRF-2010)

(A) Jamun (B) Mango

(C) Aonla (D) Grape

167. Ethylene in cold storage is removed by :

(A) Ventilation (B) UV light

(C) Low pressure (D) All of these

168. Benzoic Acid is most effective to :

(A) Mould (B) Yeast

(C) Bacteria (D) Virus

169. Potato contains following amount of protein (mg/100g of edible portion) : (JRF-1999)

(A) 1.8 (B) 1.4

(C) 2.6 (D) 1.6

170. According to FPO, the maximum limit of SO_2 allowed in squashes and cordials is :

(A) 350 ppm (B) 500 ppm

(C) 1000 ppm (D) 600 ppm

171. Optimum light for flowers in cold storage is :

(A) 100-200 ppm (B) 500-1000 ppm

(C) 1000-1500 ppm (D) 1500-3000 ppm

172. Decrease in temperatures mainly helpful to enhance shelf life of perishable by : (JRF-2007)

(A) Decreasing rate of metabolic processes

(B) Inhibiting growth of microorganisms

(C) Both (A) and (B)

(D) None of these

173. Sourness in radish is due to : (JRF-2000)

(A) Isothiosinate (B) Isosinate

(C) Anthocyanin (D) All of the above

174. Un-rooted carnation cutting can be stored at :

(A) -5°C for 2 months (B) 2°C for 2 months

(C) -5°C for 5 months (D) 2°C for 5 monts

175. Harvesting of Melons for distant market : (JRF-2010)

(A) Green mature stage (B) Half slip stage

(C) Full slip stage (D) Immature stage

176. Water disinfection is done by :

(A) Sodium Chloride (B) Potassium chloride

(C) Sodium hypochloride (D) Tween-20

177. Which of following is not commonly used as water disinfectant?

(A) UV rays (B) Aluminium Sulphate

(C) Sodium hypochloride (D) Copper sulphate

178. The constituent responsible for the aroma of apple is : (JRF-2006)

(A) Isopentanol (B) Eugenol

(C) Ethyl-2-methyl (D) Tiaxanol butyrate

179. The toxicity of SO_2 is increase at :

(A) Low temperature

(B) High temperature

(C) Moderate temperature

(D) No effect of temperature

180. Juice concentration in Fruit Jelly is : (JRF-1999)

(A) 68% (B) 40%

(C) 10% (D) 25%

181. Stable vegetable of India diet : (JRF-2010)

(A) Tomato (B) Brinjal

(C) Potato (D) Onion

182. Optimum dose of gamma radiation for prolonging vase life of flowers is :

(A) 1-5 Kra(*d*) (B) 6-10 Kra(*d*)

(C) 10-15 Kra(*d*) (D) 15-20 Kra(*d*)

183. Concentration of SO_2 in concentrated juice is :

(A) 500 ppm (B) 1000 ppm

(C) 1500 ppm (D) 350 ppm

184. Sugar concentration in Jam is : (JRF-2000)

(A) 68°Brix (B) 55°Brix

(C) 45°Brix (D) 25°Brix

185. Storage of banana done at : (JRF-2010)

(A) 11-20^{0}C & >80% RH (B) 11-20^{0}C & <70% RH

(C) 11-20^{0}C & >70% RH (D) 11-20^{0}C & <80% RH

186. Which is not suitable for dry transportation :

(A) Carnation (B) Chrysanthemum

(C) Lily (D) Iris

187. Which of the following vegetables is rich source of "Vitamin-A"? (JRF-2006)

(A) Tomato (B) Beat leaf

(C) Cucumber (D) Radish

188. Geotropic bending during transportation is common in :

(A) Gladiolus (B) Snapdragon

(C) Lupine (D) All of these

189. SO_2 content in pure KMS is :

(A) 25.5% (B) 75.5%

(C) 78.2% (D) 57.7%

190. Lactic Acid bacteria is used in : (JRF-2000)

(A) Pickle (B) Vinegar

(C) Wine (D) All of these

191. Enzyme responsible for converting pectin into pectic acid is :

(A) Pectinase

(B) Proto-peclinase

(C) Pectic Methyl Esterase (PME)

(D) Poly Galucturonase

192. ……….. is used as a preservation in the preparation of Grapes squash : (JRF-2006)

(A) KMS (B) Sodium benzoate

(C) Citric acid (D) None of these

193. Flower stems of the following is should be normally kept in boiling water for second :

(A) Rose (B) Poinsettia

(C) Lily (D) Iris

194. Yeast is used for preparing : (JRF-2000)

(A) Cider (B) RTS

(C) Pickle (D) Cordial

195. Which is flower preservative? (JRF-2010)

(A) $KMnO_4$ (B) KNO_3

(C) $HgCl_2$ (D) KMS

196. Enzyme responsible for converting protopectin into pectin is :

(A) PME (B) Proto-pectinase

(C) Poly Galucturonase (D) Pecfmase

197. CIPHET is situated at : (JRF-2000)

(A) Mysore (B) New Delhi

(C) Ludhiana (D) Bangalore

198. In home, which of following is used to increase vase life?

(A) Salt (B) Sugar

(C) Glucose (D) Oil

199. Grapes varieties which are suitable for raisin making have TSS : (JRF-2006)

(A) 12-14° Brix (B) 14-16° Brix

(C) 16-18° Brix (D) Above 19° Brix

200. The term 'climacteric' is first used by :

(A) Gane (1934)

(B) Kidd and West (1927)

(C) Cruess (1912)

(D) Bleekar (1929)

201. Staple vegetable in our diet is : (JRF-2000)

(A) Potato (B) Onion

(C) Tomato (D) Brinjal

202. The period for which flowers of a species remain presentable when kept in vase is known as :

(A) Vase life (B) Display life

(C) Shelf life (D) All of these

203. O_2 requirement for Apple storage in Controlled Atmosphere (CA) is :

(A) 2% (B) 3%

(C) 5% (D) 7%

204. Which of following fruit type is selected for Jelly making? (JRF-2005)

(A) Unripe (B) Ripe

(C) Over Ripe (D) Firm ripe

205. Refractrometer is used to measure : (JRF-2010)
(A) Acidity % (B) Viscosity
(C) TSS (D) RH

206. Storage temperature for Asparagus is :
(A) 0 - 5°C (B) 5 - 7°C
(C) 7-11°C (D) 10-15°C

207. Which of following is complex morphological unit?
(A) Fruit (B) Vegetable
(C) Seed (D) Flower

208. Which of the following crop is used against heart diseases? (JRF-2006)
(A) Senna (B) Fox glove
(C) Bellodona (D) Pyrethrum

209. Pigment which is responsible for red coloure in rose : (JRF-2000)
(A) Carotene
(B) Anthocyanins
(C) Xanthophylls
(D) Cyandin

210. Pigment which is responsible for Bluish- red coloure in rose :
(A) Carotene (B) Anthocyanins
(C) Pelargenidin (D) Cyandin

211. Salt of 8-Hydroxy Quinoline normally :
(A) Acidity water
(B) Improve water balance
(C) Act germicide
(D) All of these

212. At the time of fruit maturity specific gravity : (JRF-2010)
(A) Increase (B) Decrease
(C) No change (D) None of these

213. Pelorgonidin pigment is responsible for which of the following colour in flowers? (JRF-2007)

(A) Blue (B) Scarlet

(C) Orange (D) Red

214. Sulphur containing food is stored in : (JRF-2002)

(A) AR cane (B) SR cane

(C) Plain cane (D) All of the above

215. Which of following act as acidifying agent?

(A) Sugar (B) Citric acid

(C) Silver Nitrate (D) Cobalt Nitrate

216. Vacuum cooling is most suitable for :

(A) Fruits (B) Tubers

(C) Leafy vegetables (D) None of these

217. Oil content is maximum in : (JRF-2002, 2004)

(A) Coconut (B) Oil palm

(C) Ground nut (D) All of the above

218. Which alkaloid is not present in Sarpagandha? (JRF-2006)

(A) Solasodine (B) Serpentine

(C) Ajmaline (D) Reserpine

219. Vegetable which is not blanched before drying is :

(A) Cauliflower (B) Palak

(C) Onion (D) Tomato

220. Moisture content in dried vegetable is :

(A) 2% (B) 3%

(C) 5% (D) 6%

221. Oil content of Soybean is : (JRF-2010)

(A) 10% (B) 20%

(C) 30% (D) 40%

222. Sugar percent for effective preservation should be : (JRF-2007)
(A) 40-50 (B) 65-75
(C) 30-40 (D) 20-30

223. Bluing of rose petal is due to :
(A) Accumulation of ammonia (B) More salt
(C) Less sugar (D) Due to disease

224. The principle followed in freezing is : (JRF-2002)
(A) Crystallization (B) Drying
(C) Condensation (D) All of the above

225. Vitamin which is not found in Fruits and Vegetables is :
(A) Vitamin A (B) Vitamin B_1
(C) Vitamin B_6 (D) Vitamin B_{12}

226. Which one of the following instrument is used for the maserment of TSS? (JRF-2002, 2004)
(A) Solometer (B) Spectrophotometer
(C) Refractometer (D) Rheometer

227. Best maturity indices of orange is :
(A) TSS (B) Sugar %
(C) Acid % (D) Brix : arid ratio

228. Which one of the following is the flower preservative? (JRF-2002)
(A) $KMnO_3$ (B) KNO_3
(C) $HgCl_2$ (D) All of the above

229. Economic part of Isubgol is : (JRF-2006)
(A) Leaf (B) Seed
(C) Husk (D) Seed and Husk

230. Ayurvedic use of guggal : (JRF-2010)
(A) Reduce obesity (B) Reduce arthritis
(C) Reduce Cholesterol (D) Reduce tension

231. Bacteria which is used to absorb ethylene from storage chamber is :

(A) Agrobacterium (B) Mycobacterium
(C) Bacillus (D) Azotobacter

232. Oxalates are present in : (JRF-2002)

(A) Root crops (B) Bulb crops
(C) Leafy vegetables (D) None of the above

233. Toughening effect on canned bean is due to :

(A) K (B) Ca
(C) S (D) None of these

234. Which of following phytokinin is used for improving vase life?

(A) BA (B) IBA
(C) GA (D) PBA

235. Edible part of Safron : (JRF-2010)

(A) Petals (B) Calyx
(C) Root (D) Stigma

236. Which of following act as bactericide in vase solution?

(A) DICA (B) DDMH
(C) 8-HQC (D) All of these

237. How much quantity of colour is permitted in preservation of food? (JRF-2002)

(A) 200 ppm (B) 300 ppm
(C) 400 ppm (D) 500 ppm

238. Neera or Padaneer is a product of : (JRF-2006)

(A) Areca nut (B) Cashew nut
(C) Palmyrah palm (D) Cocoa

239. Agricultural produce (Grading and Marketing) Act (1937) is also :

(A) PFA Act (B) FPO Act
(C) Agmark Act (D) ISI Act

240. Stem plugging in cut flower is reduced by :

(A) Low pH up to 3.5 (B) Azide

(C) Calcium Nitrate (D) All of these

241. The fruits and vegetables are known as : (JRF-2002)

(A) Supplement food (B) Energy food

(C) Nutritive food (D) Protective food

242. The term "three quarterful or full three quarter" is used to denote fruit maturity in which crop in the following :

(A) Banana (B) Mango

(C) Tomato (D) Pineapple

243. Vanilla is harvested at : (JRF-2002)

(A) Tender stage (B) Fully ripe stage

(C) Ripe stage (D) Immature stage

244. ....is chemical used as anti-coagulant in processing of rubber : (JRF-2006)

(A) Formic acid

(B) Sodium sulphite

(C) Acetic acid

(D) Sulphuric acid

245. Storage of fruits and vegetables, where the gas composition is changed from that of normal atmosphere and a precise control is maintained over the atmospheric composition during storage period is known as :

(A) Controlled Atmospheric (CA) storage

(B) Modified Atmospheric (MA) storage

(C) Cold storage

(D) Hypobaric storage

246. In cordial TSS is : (JRF-2002)

(A) 20° Brix (B) 30° Brix

(C) 40° Brix (D) 50° Brix

247. During controlled atmospheric storage composition of which of the following set of gases is controlled :

(A) $O_2 + N_2$ (B) $CO_2 + N_2$

(C) $C_2H_4 + N_2$ (D) $CO_2 + O_2$

248. Fruit suitable for making jelly is : (JRF-2002)

(A) Lime (B) Apple

(C) Papaya (D) Guava

249. At which pH fruits and vegetables are divided into acidic and non-acidic for thermal processing :

(A) 4.5 (B) 5.5

(C) 6.5 (D) 7.5

250. Oranges caused chilling injury if stored below: (JRF-2006)

(A) 1°C (B) 5°C

(C) 7°C (D) 10°C

251. Maximum oil percentage is found in which of the following *Jasminum* species : (JRF-2006)

(A) *grandiflrum* (B) *auricuiatum*

(C) *sambac* (D) *Humile*

252. Aromatic plants are used for producing : (JRF-2007)

(A) Fatty oil (B) Vegetable oil

(C) Essential oil (D) Bio-diesel

253. The FPO license is given by : (JRF-2002)

(A) State Government

(B) Central Government

(C) Ministry of Agriculture

(D) ICAR

254. In pre-cooling, water is mostly removed by :

(A) Convection (B) Conduction

(C) Radiation (D) None of these

255. Which one of the following is a class one preservative? (JRF-2002)

(A) KMS (B) Sodium Benzoate

(C) Sugar (D) Silver nitrate

256. The constituent responsible for colouring of tea is :

(A) Theaflavin (B) Thearufigens

(C) Both A and B (D) None of these

257. Albinism *is* an important physiological disorder of :

(A) Plum (B) Peach

(C) Strawberry (D) Cherry

258. Sugar used as a preservative have concentration of :

(A) 40% (B) 50%

(C) 60% (D) 70%

259. Calliper grade is the maturity measurement for :

(A) Apple (B) Mango

(C) Banana (D) Pineapple

260. Bio-preservation consist the use of : (JRF-2006)

(A) Lactic acid bacteria (B) Membrane filtration

(C) Cosmic energy (D) Solar energy

261. Kottapak is the most processed product from : (JRF-2002)

(A) Cashew nut (B) Walnut

(C) Coconut (D) Areca nut

262. Formation of abscission layer is maturity index of :

(A) Tomato (B) Leafy vegetables

(C) Melons (D) Onion

263. Tetra pack pouches have : (JRF-2006)

(A) 3 layers (B) 4 layers

(C) 5 layers (D) 6 layers

264. Which one of the following is a non-climacteric fruit? (JRF-2002)
   (A) Mango (B) Banana
   (C) Pineapple (D) Guava

265. What is the maturity index for Avocado?
   (A) Sugar content (B) Acid content
   (C) TSS (D) Oil content

266. Fruit rich source of Vitamin-A is : (JRF-2002)
   (A) Apple (B) Indian gooseberry
   (C) Mango (D) Papaya

267. Which of the following is bio-degradable plastic?
   (A) Poly propylene
   (B) LDPE
   (C) Polythene
   (D) Polyhydroxy butyrate

268. The pungency of chilli is due to : (JRF-2002, 2006)
   (A) Capsacin (B) Capsanthin
   (C) Isothiocynate (D) None of the above

269. The colour of chilli is due to : (JRF-2002)
   (A) Capsacin (B) Capsanthin
   (C) Isothiocynate (D) None of the above

270. Bitterness in kinnow juice present due to : (JRF-2006)
   (A) Lutin (B) Hesperidins
   (C) Narangin (D) Tannin

271. The quick freezing is preferred over slow freezing : (JRF-2002)
   (A) Due to formation of large crystal
   (B) Due to formation of small crystal
   (C) In order to save time
   (D) Cost is less

272. As fruits mature, the specific gravity will :

(A) Increase (B) Decrease

(C) Remains constant (D) None of these

273. Sulphites and SO_2 are added to preserve fruits and vegetables products to check : (JRF-2003)

(A) Reduce acidity (B) Ethylene production

(C) Microbial activity (D) Insect attack

274. 'Solidity' is the maturity index for :

(A) Root vegetables (B) Seed vegetables

(C) Leafy vegetables (D) Cucurbits

275. Bitter principle in bitter gourd is : (JRF-2003)

(A) Capsorubin (B) Cucurbitacin

(C) Isothyosinate (D) Phenol

276. Which of the following fruits is most suitable for canning? (JRF-2006)

(A) Mango (B) Pineapple

(C) Kiwifruit (D) Banana

277. Leaf change is important maturity index for :

(A) Bulbous vegetables (B) Seed vegetables

(C) Cucurbits (D) Leafy vegetables

278. Which one of the following is not a food preservative? (JRF-2004)

(A) KMS (B) Sodium benzoate

(C) Benlate (D) Glacial acetic acid

279. Mango fruits can be best stored at a temperature of :

(A) 8°C (B) 16°C

(C) -4°C (D) 0°C

280. One tone flowers of jasmine, how much absolute oil is extracted?

(A) 0.5-1.0 Kg (B) 2.5-3.0 Kg

(C) 1.3-1.5 Kg (D) 2 Kg

281. Which of the following alkaloids from Datura is used in the relief of withdrawal symptoms of morphine and alcohol addition? (JRF-2004)

(A) Hyoscine (B) Atropine

(C) Meteloidine (D) Hyoscyamine

282. Seed of which of the following are used in extraction of Jasmine oil?

(A) Caster (B) Mustard

(C) Sesame (D) Ground nut

283. Which of the following plant hormone is considered as ripen :

(A) Cytokinin (B) GA_3

(C) Ethylene (D) IAA

284. The oil content is more in Damask rose flower produce at :

(A) Lucknow (B) Ajmer

(C) Delhi (D) Palampur

285. "Protective food" includes : (JRF-2004)

(A) Fruits & vegetables (B) Cereals

(C) Pulses (D) Fruits

286. Bitterness in Citrus juice is due to :

(A) Sugar (B) Acid

(C) Glucosides (D) Vitamins

287. The concrete content in carnation flowers is about :

(A) 1.0-2.0% (B) 2.0-3.0%

(C) 0.2-0.3% (D) 0.02-0.03%

288. Richest source of Vitamin-C is : (JRF-2004)

(A) Aonla (B) Guava

(C) Citrus (D) Bael

289. Who is the largest importer of indoor plants? (JRF-2006)

(A) Italy (B) USA

(C) Netherland (D) Germany

290. Bitterness in peach is due to :

(A) Sugar (B) Malic acid

(C) Hydrocyanin (D) Prunasin acid

291. Marmalade is prepared from the fruits of : (JRF-2004, 2009)

(A) Apple (B) Guava

(C) Citrus (D) Mango

292. Which country is the largest producer is carnation perfume?

(A) India (B) China

(C) USA (D) France

293. Among the following, which is best maturity index for Grape :

(A) Size (B) Shape

(C) Colour (D) TSS

294. For long vase life optimum CO_2 concentration should be in green house :

(A) 500 ppm (B) 1000 ppm

(C) 1500 ppm (D) 3000 ppm

295. Which country is the largest producer in *Michelia champacan* perfume?

(A) India (B) China

(C) USA (D) Egypt

296. Which of following constitute about 40% of total cut foliage in global trade?

(A) Asparagus (B) Leather leaf fern

(C) Podocarpus (D) Eucalyptus

297. Which one is a ripening hormone? (JRF-2004)

(A) Auxin (B) Ethylene

(C) GA_3 (D) All of these

298. Which salt of following is not a determine to flower longevity?

(A) Chlorine (B) Fluorine

(C) Aluminium (D) Calcium

299. Toddy from coconut is prepared by :

(A) Drying (B) Freezing

(C) Dehydration (D) Fermentation

300. Which is a non-climacteric fruit? (JRF-2004)

(A) Jackfruit (B) Banana

(C) Orange (D) Papaya

301. Which of following accelerate senescence in cut carnation?

(A) Abscissic acid (B) Ethylene

(C) High temperature (D) All of these

302. A lip opposite to odd sepal in orchid is known as :

(A) Colmn (B) Labellum

(C) Kell (D) None of these

303. Which species of *Tagetes* is commercially grown for oil extraction :

(A) *patula* (B) *Minuta*

(C) *erecta* (D) *Tenuifolia*

304. Short days at 1-2 leaf stages in gladiolus leads to :

(A) Bud Blasting (B) Blind shoot

(C) Poor spike (D) Multiple-spike

305. Which of following is a non-climacteric fruit? (JRF-2006)

(A) Litchi (B) Mango

(C) Apple (D) Banana

306. Which is a climacteric fruit? (JRF-2004)

(A) Bael (B) Grapes

(C) Pomegranate (D) Litchi

307. Maximum density of water is at a temperature of :

(A) 0°C (B) 4°C

(C) 4°C (D) -7°C

308. Ancymidol is used in which plant for to reduce elongation of last inter-node :

(A) Lily (B) Tulip

(C) Daffodil (D) Carnation

309. Which of following preservative show phytotoxicity in daffodil :

(A) 8-HQC (B) STS

(C) Citric acid (D) Nickel Chloride

310. Sweetest sugar in fruits is : (JRF-2004)

(A) Glucose (B) Sucrose

(C) Fructose (D) Galactose

311. The chemical used for preservation of fruit products is : (JRF-2009)

(A) Potassium permanganate

(B) Potassium metabisulphite

(C) Potassium sulphate

(D) All of the above

312. Bio-preservation consiste the use of : (JRF-2006)

(A) Lactic acid bacteria (B) Solar energy

(C) Membrane filtration (D) Cosmic energy

313. Apinasty in poinsettia can be reduce by spraying :

(A) Sugar (B) Silver Nitrate

(C) Cobalt Nitrate (D) Ancymidol

314. Guava fruit is botanically known as :

(A) Drupe (B) Sorosis

(C) Berry (D) Pome

315. Optimum CO_2 concentration for production of potted Geranium is :

(A) 100-300 ppm (B) 300-500 ppm

(C) 500-700 ppm (D) 700-1000 ppm

316. Waxing of fruit is done mainly to reduce : (JRF 2004)
   (A) Respiration
   (B) Transpiration
   (C) Ripening
   (D) A and B both

317. Mostly dry fruit are rich in : (JRF-2005, 2006)
   (A) Protein (B) Carbohydrates
   (C) Fats (D) Vitamins

318. Who collected drops of rose oil from canal filled with rose petal and named it 'Attre Jahangiri'?
   (A) Shahjahan (B) Noorjahan
   (C) Mumtaj (D) Noormahal

319. Which of the following is not a *Rosa* Species grown for oil extraction?
   (A) *Indica* (B) *Damascene*
   (C) *Bourboniana* (D) *Centifolia*

320. Which of the following is a rich source of fat : (JRF-2000, 2004, 2006)
   (A) Cashew nut (B) Walnut
   (C) Almond (D) Avocado

321. Which country is the largest producer of rose oil?
   (A) Bulgaria (B) Italy
   (C) India (D) China

322. Chilling injury is common in which orchid species :
   (A) Cattleya (B) Cymbidium
   (C) Venda (D) Paphiopedilum

323. Red colour of tomato is due to : (JRF-2008)
   (A) Anthocyanin (B) Xanthophyll
   (C) Lycopene (D) Carotene

324. The heat treatment given to inactive enzymes is called : (JRF-2004)

(A) Cooling (B) Oxidation

(C) Drying (D) Blanching

325. Blue colour of flower petal is due to : (JRF-2006, 2008)

(A) Delphinidin (B) Cynidin

(C) Pelargonidin (D) Beta-carotene

326. In onion pink colour is due to :

(A) Anthocyanin (B) Carotene

(C) Xanthophyll (D) Quercitin

327. Lye peeling is done by : (JRF-2004, 2009)

(A) Sodium hydroxide

(B) Potassium hydroxide

(C) Potassium sulphate

(D) Calcium sulphate

328. One Kg rose oil is extracted from :

(A) 10 q flowers (B) 3-5 ton flowers

(C) 1.5 ton flowers (D) 100 Kg flowers

329. Zero energy cool chamber is developed by :

(A) M K Rai and RN Singh

(B) SK Roy and DS Khurdiya

(C) RP Roy and DK Khurana

(D) None of these

330. The average daily requirement of vegetable for human adult is : (JRF-2004, 2006)

(A) 120g (B) 220g

(C) 285g (D) 320g

331. In which state of India distillation of rose is done :

(A) Punjab (B) UP

(C) Rajasthan (D) All of these

332. Deficiency of "Vitamin-C" in human diet lead to the development of : (JRF-2004)

(A) Scurvy (B) Beri-Beri

(C) Pellagra (D) Goitrel

333. Pellagra is due to deficiency of : (JRF-2009)

(A) Folic acid (B) Pyridoxine

(C) Niacin (D) Biotin

334. Which of following is species of *Jasminum* grown for oil extraction :

(A) *sambac* (B) *grandiflorum*

(C) *auriculatum* (D) All of these

335. The membrane lipid hypothesis is given by :

(A) Kidd & West

(B) James Harrison

(C) Raison & Lyons

(D) Graham & Patterson

336. Jasmine distillation units are commercially setup in state of :

(A) Punjab (B) Karnataka

(C) TN (D) UP

337. The main process involved in the post harvest losses of fruits and vegetables are : (JRF-2007)

(A) Transpiration (B) Respiration

(C) Photosynthesis (D) Both (A) and (B)

338. The point at which the dried products just become lumpy is known as :

(A) Danger Point (B) Saturated Point

(C) Critical Point (D) Safety Point

339. Roots of which of the following use for oil extraction? (JRF-2011)

(A) Rose (B) Sweet pea

(C) Gardenia (D) Iris

340. Which of the following fruit has maximum protein(21%) : (JRF-2004)

(A) Cashew nut (B) Walnut

(C) Almond (D) Avocado

341. Which country is the largest producer of Jasmine oil?

(A) India (B) USA

(C) Bulgaria (D) Egypt

342. A Point which have 5% Low RH than the Critical Point is :

(A) Critical Point (B) Danger Point

(C) Saturated Point (D) Safety Point

343. Edible part of pomegranate : (JRF-2005)

(A) Fleshy arils (B) Recipectels

(C) Fleshy thallamus (D) Cotyledons

344. Which species of Jasminum is grown for high quality perfume?

(A) *sambac* (B) *grandiflorum*

(C) *flexile* (D) *multiflorum*

345. Which of following is most common solvent in perfumery?

(A) Distilled Water (B) Acetone

(C) Petroleum ether (D) Tarpene

346. Fungus which mostly grown on grapes :

(A) *Geotrichum* (B) *Penicillium*

(C) *Botrytis* (D) *Colletotrichum*

347. Which of the following flowers is most suitable for drying? (JRF-2007)

(A) Poppy (B) Acroclinum

(C) Tuberose (D) Cosmos

348. "Vitamin-D" is chemically known as:

(A) Retinol (B) Cabalanin

(C) Calciferol (D) Tocopherol

349. Cider is preserved by : (JRF-2005)

(A) Benzoic acid (B) Salt

(C) Sugar (D) SO_2

350. Lye peeling is done at a temperature of :

(A) 75°C (B) 84°C

(C) 93°C (D) 105°C

351. Pre-cooling is meant to : (JRF-2005, 2009)

(A) Remove field heat

(B) Improve fruit quality

(C) For remove of dart

(D) All of the above

352. Banana chips are prepared best from variety : (JRF-2005)

(A) Cavandish (B) Nandran

(C) Rasthali (D) Riobrusta

353. Which colour is considered as warm colour :

(A) Blue (B) Orange

(C) Green (D) Violet

354. Pectin content is important for the preparation of : (JRF-2005)

(A) Jam (B) Jelly

(C) Ketchup (D) Marmalade

355. Oil content used to measure harvest maturity : (JRF-2009)

(A) Oil palm (B) Date palm

(C) Avocado (D) Karondha

356. What is the threshold level of ethylene in fruit and vegetable :

(A) 0.01 μL/L (B) 0.02 μL/L

(C) 0.03 μL/L (D) 0.04 μL/L

357. Protein is found in abundance in : (JRF-2005)

(A) Carrot (B) Bitter gourd

(C) Chilli (D) Beans

358. Which of the following is a rapid pre-cooling method :

(A) Forced air Cooling (B) Hydro Cooling

(C) Vacuum Cooling (D) Evaporative Cooling

359. Which of the following shrub is grown for oil extraction?

(A) *Gardenia* (B) *Hibiscus*

(C) *Cestrum* (D) *Sambucus*

360. Tomato is rich source of : (JRF-2009)

(A) Iron (B) Vitamin A

(C) Vitamin-C (D) Phosphorus

361. Optimum storage temperature and relative humidity range for apple are : (JRF-2009)

(A) -3^0 to -1^0C and 85-90% RH

(B) -1^0 to 0^0C and 90-95% RH

(C) 0^0 to 1^0 C and 92-96% RH

(D) 0.5^0 to 1.5^0C and 93-98% RH

362. Which of the following is a ethylene absorbent?

(A) K_2SO_4 (B) KCl

(C) $KMnO_4$ (D) KNO_3

363. The safe seed storage moisture control should be between : (JRF-2009)

(A) 4-5% (B) 8-9%

(C) 13-14% (D) 15-16%

364. Which of the following annual is commercially grown for oil extraction :

(A) Pansy (B) Paper flower

(C) Salivia (D) Sweet pea

365. Cassava is used for the manufacture of : (JRF-2005)

(A) Latin (B) Sago

(C) Pectin (D) Sugar

366. In cucumber, chilling- injury symptoms are occurred at :
(A) <7°C (B) 7°C
(C) 10°C (D) >10°C

367. Sauerkraut can be made from : (JRF-2009)
(A) Broccoli (B) Cabbage
(C) Chinese cabbage (D) Spinach

368. Which country is the largest producer of cut flowers in the world?
(A) Netherlands (B) Germany
(C) Spain (D) USA

369. Bitterness in kinnow juice present due to: (JRF-2006)
(A) Citric acid (B) Limonene
(C) Hydrocyanin (D) Narangin

370. De-greening is not applicable in :
(A) Banana (B) Guava
(C) Mango (D) Citrus

371. Which of the following is largest importer of dry flowers?
(A) USA (B) UK
(C) India (D) Germany

372. Delphinidin a plant pigment is responsible for development of .... colour. (JRF-2009)
(A) Blue (B) Orange
(C) Red (D) Pink

373. Carrot is a good source of : (JRF-2005)
(A) Vit-A (B) Vit-B
(C) Vit-C (D) Protein

374. Deficiency of which of the following vitamin cause scurvy disease? (JRF-2007)
(A) Vit-A (B) Vit-B
(C) Vit-C (D) Vit-D

375. Under normal conditions Orchid can be stored upto 2 weeks at :

(A) 2-4°C (B) 5-7°C

(C) 10-12°C (D) 1°C

376. Mango pulp is preserved by : (JRF-2005)

(A) Citric acid (B) KMS

(C) Sodium benzoate (D) Sugar

377. In spray drying method, the food material used, is :

(A) Solid (B) Gas

(C) Lliquid (D)

378. Fumigation is the process to kill :

(A) Insects (B) Eggs

(C) Both Insects and Eggs (D)

379. In cryogenic freezing process, the temperature is below :

(A) -10°C (B) -20°C

(C) -60°C (D) -30°C

380. The process which is not coming under canning :

(A) Hot Packing (B) Freezing

(C) Pasteurization (D) Washing

381. Pre-harvest spray of which the following chemical is done to reduce storage loss of onion : (JRF-2006)

(A) IBA (B) Maleic Hydrazide

(C) GA_3 (D) Ethylene

382. Gram contains about............percent fat. (JRF-2007)

(A) 3.5 (B) 4.5

(C) 6.5 (D) 7.5

383. Which is the largest fruit in the world?

(A) Water Melon (B) Bottle gourd

(C) Jack Fruit (D) Avocado

384. Which of the following is exporter of dry flowers :
(A) India (B) Australia
(C) South Africa (D) All of these

385. 'Curcumin' is extracted from :
(A) Turmeric (B) Chilli
(C) Black pepper (D) Clove

386. Which of following flowers grown for dry flowers?
(A) Strew flower (B) Statice
(C) Paper flower (D) All of these

387. Important tree vegetable of perennial nature is : (IARI-2012)
(A) *Murraya keonigii*
(B) *Capsicum annum*
(C) *Vigna unguiculata*
(D) *Phaseolus vulgaris*

388. Kwashhiorkor is due to the deficiency of : (IARI-2012)
(A) Vitamin-A (B) Vitamin-B
(C) Vitamin-C (D) Protein

389. Indian export of dry flowers constitute about :
(A) 10-15% (B) 20-30%
(C) 50-60% (D) 60-70%

390. Optimum RH for storase of onion and garlic :(IARI-2012)
(A) 50-60% (B) 61-70%
(C) 71-80% (D) 81-90%

391. Father of quick freezing is : (IARI-2012)
(A) Kidd and West (B) S.P. Burg
(C) S.Ranganna (D) Clarence Birdseye

392. The main dry flower product been exported from India is :
(A) Lotus pods (B) Hybrid lily
(C) Rose (D) Carnation

393. Which is the largest dry flowers exporting company from India?
(A) Deccan Flora (B) Kasturi Flowers Ltd.
(C) Ramesh Flower Ltd. (D) Cosco Flora

394. Proteolytic enzyme in pineapple is : (IARI-2012)
(A) Bromelin (B) Diastase
(C) Ficin (D) Papain

395. Most of the companies exporting dry flowers are situated at :
(A) Bangalore (B) Delhi
(C) Mumbai (D) Tuticorin

396. Black neck in tomato ketchup is due to the : (IARI-2012)
(A) Ferric tannate (B) Sodium benzoate
(C) Ferric sulphate (D) Acetic acid

397. Acidic fruit juices are sterilized under : (IARI-2012)
(A) Low pressure
(B) High pressure
(C) Normal atmospheric pressure
(D) Vacuum

398. Folic acid rich vegetable is : (IARI-2012)
(A) Spinach (B) Pulses
(C) Lettuce (D) Brussels's sprouts

399. Which is a famous product of dry flowers is very common in Europe?
(A) Wall Sceneries (B) Potpourris
(C) Pot plants (D) Flower arrangement

400. Most important pathogenic bacteria causing spoilage of food is : (IARI-2012)
(A) *Clostridium botulinum*
(B) *Bacillus cereus*
(C) *Streptococcus aureus*
(D) *Salmoneila typhimurium*

401. Blanching is done to :(IARI-2012)

(A) Inactivate enzymes

(B) Remove the peel

(C) Remove moisture

(D) Remove microorganisms

402. The means to dry something under artificially produced heat through controlled temperature, humidity and air flow is known as :

(A) Cooling (B) Dehydration

(C) Desiccation (D) None of these

403. Vinegar is : (IARI-2012)

(A) 1% acetic acid (B) 10% acitic acid

(C) 40 grain acetic acid (D) 80 grain acetic acid

404. Potassium permanganate is used in sealed packets to : (IARI-2012)

(A) Cerate an inert atmosphere

(B) Reduce the level of CO_2

(C) Reduce the level of ethylene

(D) Increase O_2 level

405. Which of the following has beautiful natural dried fruits?

(A) Amaltas (B) Bael

(C) Pines (D) All of these

406. Freezing preservation involves the principle of : (IARI-2012)

(A) Addition of cold (B) Removal of heat

(C) Addition of heat (D) Addition of ice

407. Which of the following organism produces single cell protein : (IARI-2012)

(A) *Methylophilus methyiotrophus*

(B) *Salmonellia* sp.

(C) *Trichinella*

(D) *E.coli*

408. The ideal conditions for ripening of mango are : (IARI-2012)
(A) 0-2°C and 80-95% RH
(B) 2-4°C and 80-90% RH
(C) 18-25°C and 85-90% RH
(D) 30-35°C and 80-85% RH

409. Which is beautiful natural dried seed?
(A) Indian liquorice (B) Hedera
(C) Marigold (D) Wood rose

410. The important quality parameter for export of dehydrated onion is:(IARI-2012)
(A) Red colour with low pungency
(B) White colour with low pungency
(C) Brownish yellow colour
(D) Redish brown colour

411. Spectrophotometer with visible wavelength should have range between : (IARI-2012)
(A) 300-500 nm (B) 380-760 nm
(C) 480-900 nm (D) 500-960 nm

412. Which of the following is most suitable for air drying?
(A) Paper flower (B) Anaphalis
(C) Straw flower (D) All of these

413. The number of ATP molecules synthesized on complete oxidation of a hexose sugar would be : (IARI-2012)
(A) 20 (B) 32
(C) 38 (D) 56

414. Any material which remove moisture without reacting with it is known as :
(A) Desiccant (B) Dehydration
(C) Absorbent (D) Drier

415. For canning factory analysis is important for : (IARI-2012)
   - (A) Sensory quality of food material
   - (B) Packaging of row material
   - (C) Storage quality of food material
   - (D) Bacteriological quality of water

416. The desiccant should have an ideal size of :
   - (A) 1-2 mm
   - (B) 0.02-0.2 mm
   - (C) 2-5 mm
   - (D) All of these

417. Waxing of fruits is done mainly to reduce : (IARI-2012)
   - (A) Transpiration
   - (B) Respiration
   - (C) Ripening
   - (D) Transpiration and respiration

418. The major indicator and a useful guide to perdict the potential storage life of the fruits and vegetables is : (IARI-2012)
   - (A) Water uptake rate
   - (B) Rate of ascorbic acid loss
   - (C) Rate of mineral uptake
   - (D) Respiration rate

419. The most accepted pathway of ethylene biosynthesis is : (IARI-2012)
   - (A) β-alanine pathway
   - (B) Ethanol Pathway
   - (C) Methionone pathway
   - (D) Linolenic acid pathway

420. Which of the following is suitable desiccant for flower drying?
   - (A) Silica gel
   - (B) Sand
   - (C) Boric acid
   - (D) All of these

421. Enzyme responsible for conversion of sucrose to glucose and fructose during respiration is : (IARI-2012)

(A) Hexokinase (B) Maltase

(C) Dehydrogenase (D) Invertase

422. In hot air oven which is the optimum temperature for drying of flowers :

(A) 30-40°C (B) 40-60°C

(C) 60-80°C (D) 80-100°C

423. Optimum harvest maturity of Gerbera flower is decided : (IARI-2012)

(A) When the flower is fully open

(B) When centre of the oldest floret opens

(C) When the flower is half open

(D) When outer rows show pollens

424. In general, longevity of cut flowers is increase with : (IARI-2012)

(A) Increase lipid level

(B) Decreased lipid level

(C) Increased carbohydrate level

(D) Decreased carbohydrate level

425. Which of the following is the quickest method of flower drying?

(A) Microwave (B) Hot air oven

(C) Solar dryer (D) Room drying

426. For better post harvest quality, the cut flowers should be harvested during : (IARI-2012)

(A) Evening hours (B) Early morning hours

(C) Noon time (D) Late in the night

427. A visible shrinkage of freshly harvested fruits and vegetables appear when the moisture loss is around : (IARI-2012)

(A) 1-2% (B) 2-3%

(C) 3-5% (D) 10-15%

428. Lye peeling is a technique of peeling of fruits and vegetables using : (IARI-2012)

(A) 1-2% HCl (B) 1-2% H_2SO_4

(C) 1-2% NaOH (D) 1-2% KOH

429. ....drying is done at very low temperature up to -35°C :

(A) Vacuum (B) Freeze

(C) Cool (D) Air

430. 'Exhausting' a term in canning is associated with : (IARI-2012)

(A) Removal of microbes (B) Removal of air

(C) Removal of flavour (D) Removal of colour

431. A dividing line of pH between acid and non-acid foods is : (IARI-2012)

(A) pH= 3.5 (B) pH = 4.5

(C) pH= 5.0 (D) pH= 5.5

432. Which of the following are dried by water drying method?

(A) Hydrangea (B) Corn flower

(C) Baby's breath (D) All of these

433. The time-temperature combination which has been received by a food during sterilization is usually expressed as : (IARI-2012)

(A) D-value (B) Z-value

(C) F-value (D) All of the above

434. Which is dried flower are used for making greeting cards?

(A) Air (B) Embedded

(C) Press (D) Freeze

435. Which one of the following driers takes least time for dryer : (IARI-2012)

(A) Hot air drying

(B) Heated surface drier

(C) Infra-red drier

(D) Microwave drier

436. 'Hedonic scale' is used to determine : (IARI-2012)

(A) Physical quality

(B) Chemical quality

(C) Sensory quality

(D) Microbiological quality

437. In food preservation most commonly used irradiation is : (IARI-2012)

(A) X-rays (B) α-rays

(C) β-rays (D) γ- rays

438. The natural red colour pigment which changes its colour with pH is : (IARI-2012)

(A) Carotene (B) Xanthophyll

(C) Betalain (D) Anthocyanin

439. Which of the following is not a serious pest of dry flowers?

(A) Mice (B) Silver fish

(C) Book lice (D) Caterpillars

440. Fruit juices may be clarified to get rid of very minute particles and colloidal materials by : (IARI-2012)

(A) Filtration (B) Enzymes

(C) Centrifugation (D) Boiling

441. Deaeration in orange juice is done primarily to prevent : (IARI-2012)

(A) Destruction of vitamin-A

(B) Destruction of vitamin-C

(C) Destruction of vitamin-E

(D) All of the above

442. In any chromatography two thing are common in all : (IARI-2012)

(A) Solvent and column (B) Solvent and detector

(C) Stationary and mobile phase (D) Detector and injector

443. Acid resistant tin cans are also called : (IARI-2012)

(A) C-enamel (B) L-enamel

(C) R-enamel (D) S-R enamel

444. Sauerkraut means : (IARI-2012)

(A) Acid cabbage (B) Sweet cabbage

(C) Blanched cabbage (D) Dehydrated cabbage

445. For preparing fruit jelly, quantity of sugar to be added depends upon : (IARI-2012)

(A) Protopectin (B) Starch

(C) Pectin (D) Pectin acid

446. Oxidation of fats in potato chips is prevented by : (IARI-2012)

(A) Preservative (B) N_2 gas packaging

(C) Citric acid (D) Colouring agent

447. The browning of fresh potato cut surface is due to : (IARI-2012)

(A) Non-enzxymatic browning

(B) Mallard reaction

(C) Fermentation

(D) Enzymatic browning

448. Formation of brown mass due to burning of dry sugar is called as: (IARI-2012)

(A) Charcoal (B) Caramel

(C) Browning (D) None of these

449. Mixing of 1 kg of sugar in 1 kg of water gives sugar syrup of °Brix : (IARI-2012)

(A) 25 (B) 50

(C) 75 (D) 100

450. The most important commercial product prepared from the grapes is : (IARI-2012)

(A) Pickle (B) Raisine

(C) Marmalade (D) Wine

451. The Food Safety Act was issued by the MOFPI in the year : (IARI-2012)

(A) 2001 (B) 2003

(C) 2005 (D) 2006

452. Specific gravity is the ratio of : (IARI-2012)

(A) Weight/volume

(B) Volume/Weight

(C) Weight/volume X 100

(D) Volume/Weight X 100

453. Convering of candied fruits with a thin transparent coating of sugar which imparts them glossy appearance is known as : (IARI-2012)

(A) Blending (B) Baking

(C) Glazing (D) Syruping

454. TSS is maximum in : (IARI-2012)

(A) RTS drink (B) Squash

(C) Natural fruit juices (D) Jam

455. Spongy tissue in mango fruit is due to : (IARI-2012)

(A) Disease

(B) Physiological disorder

(C) Mechanicaldamage

(D) Chiling injury

456. Which of the following fruits is rich in riboflavin : (IARI-2012)

(A) Apple (B) Banana

(C) Guava (D) Bael fruit

457. Vacuum cooling is ideally suitable : (IARI-2012)

(A) Root vegetable

(B) Temperate fruit

(C) Leafy vegetable

(D) Fruit type vegetable

458. Which of the following fruit belongs to climacteric group : (IARI-2012)

(A) Pomegranate (B) Passion fruit
(C) Litchi (D) Loquat

459. Which of the following fruits is stored commercially under CA condition :(IARI-2012)

(A) Papaya (B) Apple
(C) Lemon (D) Pineapple

460. Who is considered as the 'Father of the canning industry' : (IARI-2012)

(A) Peter Durand (B) Metalmikoft
(C) Nicholas Appaert (D) Thomas Sadditington

461. Albedo of citrus fruits is rich in : (IARI-2012)

(A) Acid (B) Sugar
(C) Pectin (D) Colour

462. The 'Flate sour' spoilage causing bacteria of canned foods are : (IARI-2012)

(A) Mesophilic (B) Psychrophilic
(C) Thermophilic (D) Halophilic

463. Green portion of potato is undesirable as it may contain harmful substances like: (IARI-2012)

(A) Chlorophyll (B) Amylase
(C) Solanin (D) Lipoproteins

464. Name the lowest a_w value for permitting the growth of osmophillic yeasts : (IARI-2012)

(A) 0.91 (B) 0.88
(C) 0.75 (D) 0.60

465. Degreening of citrus fruit is done by the application of : (IARI-2012)

(A) O_2 (B) Ethylene
(C) CO_2 (D) Nitrogen

466. Easily useful fermentable component in food is : (IARI-2012)

(A) Fiber (B) Carbohydrate

(C) Lipid (D) Protein

467. Under commercial practice, the recommended ethylene dose for ripening of banana is : (IARI-2012)

(A) 1000 ppm (B) 100 ppm

(C) 10 ppm (D) <10

468. Symptoms associated with chilling injury are : (IARI-2012)

(A) Freezing injury and burn

(B) Freezing burn and discoloration

(C) Internal discoloration and electrolytic leakage

(D) Electrolytic leakage and burning

469. Laminated film with lowest water vapour and oxygen transmission rate is : (IARI-2012)

(A) Saran (B) OPP

(C) PE (D) CPP

470. Pro-vitamin A activity is found in : (IARI-2012)

(A) Lycopen (B) β-carotene

(C) Xanthophyll (D) Lutein

471. The compound which is common in Aonla, Jamun and Tea : (IARI-2012)

(A) Ascorbic acid (B) Organic acid

(C) Amino acid (D) Phenols

472. Lemon squesh is generally preserved by : (IARI-2012)

(A) Acetic acid (B) Citric acid

(C) Sugar (D) KMS

473. Which bean is used for extruction of gum :

(A) Cluster bean (B) Broad bean

(C) French bean (D) Hyacinth bean

474. Chemical used for removing the spray residues of Arsenic acid and lead from the fruit :

(A) H_2SO_4 (B) HCl

(C) NaCl (D) NaOH

475. Enzyme responsible for converting pectin into pectic acid is :

(A) Poly Galucturonase (PG)

(B) Pectic Methyl Esterase (PME)

(C) Proto-pectinase

(D) Pectinase

476. Enzyme responsible for converting protopectin into pectin is :

(A) PME (B) PG

(C) Proto-pectinase (D) Pectinase

477. Aroma in Guava is due to :

(A) Methyl butyrate

(B) Methyl anthranilate

(C) Estragole

(D) β-caryophyllene

478. Non reducing sugar is :

(A) Sucrose (B) Fructose

(C) Glucose (D) Mannose

479. Citric acid is commercially produced by :

(A) *Bacillus* (B) *Penicillium*

(C) *Aspergillus niger* (D) *Clostrium*

480. Public Health Laboratory is situated at :

(A) Pune (B) Kolkata

(C) New Delhi (D) Mysore

481. Central Food Laboratoryis located at :

(A) New Delhi (B) Kolkata

(C) Mumbai (D) Mysore

482. TSS of bael pulp is :

(A) 15 °Brix (B) 20 °Brix

(C) 25 °Brix (D) 30 °Brix

483. Moisture content in Intermediate Moisture Food (IMF) is :

(A) 20-30% (B) 30-40%

(C) 40-50% (D) 50-60%

484. Water activity (a_w) of mostly Intermediate Moisture Food (IMF) is :

(A) 0.25-0.50 (B) 0.50-0.75

(C) 0.75-0.85 (D) 0.85-1.0

485. The pigment present in Jamun is :

(A) Anthocyanin (B) Cartenoid

(C) Lycopens (D) Xanthophyll

486. Which one of the following fruit has maximum export from India to foreign countries?

(A) Banana (B) Citrus

(C) Apple (D) Mango

487. Which one of the following fruit product has maximum export form India?

(A) Lime Pickel (B) Mango Pickel

(C) Mango Pulp (D) Cashewnut

488. The major mineral, which is present in fruit and vegetables is :

(A) Na (B) Ca

(C) K (D) Mg

489. Browning in apple is due to :

(A) Poly Phenol oxidase

(B) Maillard reaction

(C) Carmelization

(D) Ascorbic Acid Browning

490. The 'Fruit Product Order' (FPO) licence is given by :
   (A) ICAR
   (B) Govt. of India
   (C) State Govt.
   (D) Ministry of Agriculture

491. Bitter principle in bittergourd is :
   (A) Terpenoids (B) Cucurbitacin
   (C) Glucosinolates (D) Calcium oxalate

492. A separate ministry of Food Processing was establish in :
   (A) 1954 (B) 1955
   (C) 1988 (D) 1968

493. Bitterness in Citrus juice is due to :
   (A) Glucosides (B) Acid
   (C) Preservative (D) All of these

494. Bitterness in Peach is due to presence to :
   (A) Oxalic acid (B) Citric acid
   (C) Tartaric acid (D) Prunasin acid

495. Aroma in coffee is due to :
   (A) Drying (B) Grinding
   (C) Roasting (D) Fermentation

496. Maximum density of water is at a temperature of :
   (A) -4°C (B) 0°C
   (C) 4°C (D) 7°C

497. Pungency in radish is due to presence of :
   (A) Anthocyanins (B) Glucosinolates
   (C) Terpenoids (D) Isothiosinate

498. Which stimulant is present in Tea :
   (A) Caffein (B) Thiobromin
   (C) Morphine (D) None of these

499. Cryo-preservation is related to :

(A) Liquid Oxygen (B) Liquid Nitrogen

(C) Liquid Ammonia (D) Liquid CO_2

500. In cashew apple, edible portion is :

(A) Kernel (B) Pericarp

(C) Peduncle (D) Endosperm

501. What should be the ratio of acid and sugar in good jelly?

(A) 65:1 (B) 1:65

(C) 1:55 (D) 55:1

502. Which of the organization is engaged in exporting of processed products?

(A) NAFED (B) NABARD

(C) FCI (D) APEDA

503. Organization which is related to agriculture marketing :

(A) NAFED (B) NABARD

(C) FCI (D) APEDA

504. Betalians is a active pigment which is found in :

(A) Carrot (B) Beet root

(C) Tomato (D) Radish

505. First Indian Horticulture Congress was held in :

(A) 1942 (B) 1954

(C) 1988 (D) 2004

506. The Horticulture Society of India was established in :

(A) 1942 (B) 1954

(C) 1955 (D) 2004

507. Bromelin is present in :

(A) Banana (B) Pineapple

(C) Apple (D) Mango

508. Brinjal proteccts against the :
(A) Cancer (B) Allergy
(C) Diabetes (D) Bery-Bery

509. Active compound present in Bael is :
(A) Ficin (B) Marmelosin
(C) Annocin (D) Quartecin

510. Kiwi fruit is also known as :
(A) Chinese Gooseberry
(B) Indian Gooseberry
(C) Strawberry
(D) Blackberry

511. Which of the following is a Intermediate Moisture Food (IMF)?
(A) Pickle (B) Cordial
(C) Paste (D) Jelly

512. Which enzyme is used for clarification of fruit juices?
(A) Lipoxygenase (B) Actinidin
(C) Pectinol (D) Bromelain

513. Water Activity (a_w) for tomato paste is :
(A) >0.98 (B) <0.6
(C) 0.93-0.98 (D) 0.83-0.93

514. Equilibrium Relative Humidity (ERH)of pure water is :
(A) 25% (B) 50%
(C) 75% (D) 100%

515. Water activity (aw) of pure water is :
(A) 0.25 (B) 0.50
(C) 0.75 (D) 1.00

516. Black neck is related to :
(A) Rose (B) Carnation
(C) Sauce (D) Pickle

517. In India, processing industries utilizing of total fruit production :

(A) >1% (B) <1%

(C) 1% (D) 0.1%

518. The optimum temperature for growth of thermophic bacteria is :

(A) 25°C (B) 35°C

(C) 45°C (D) 55°C

519. At which, a_w non-enzymatic browning start in dried foods :

(A) 0.5 (B) 0.6

(C) 0.7 (D) 0.8

520. The first example of the use of bio-agent for controlling post harvest diseases is the use of :

(A) *Candida oleophila*

(B) *Pseudomonas syrigae*

(C) *Bacillus subtilis*

(D) *Trichoderma harzianum*

521. The moisture content of dried cashewnuts should be :

(A) Nearly 2% (B) Nearly 5%

(C) Nearly 8% (D) Nearly 12%

522. Button mushrooms are harvested when the size of their cap is between :

(A) 10-20mm (B) 20-30mm

(C) 30-40mm (D) 40-50mm

523. Blue mold of apple is caused by :

(A) *Penicillium expansum*

(B) *Penicillium digitatum*

(C) *Penicillium italicum*

(D) *Botrytis cinerea*

524. Strawberry fruit is derived from which plant tissue?

(A) Pericarp (B) Receptacle

(C) Peddicle (D) Peduncle

525. Carotenoides in plant cell are stored in :

(A) Amyloplasts (B) Chloroplasts

(C) Chromoplasts (D) Ribosomes

526. Distinctive flavor in Jamun is due to the presence of :

(A) Methyl propionate

(B) Methyl isothiocyanate

(C) Methyl anthranilate

(D) Hexal

527. AH, B theory of sweet taste reception was proposed by :

(A) Solms (1971)

(B) Shallenberger and Acree (1969)

(C) Marse (1981)

(D) Ikeda (1909)

528. Umami is related to :

(A) Glutamic acid (B) Tartaric acid

(C) Gymnemic acid (D) Ascorbic acid

529. Strawberry contains a acid, which has anti-cancerous properties :

(A) Gallic acid (B) Ellagic acid

(C) Phytic acid (D) Citric acid

530. Celiae' disease is associated with consumption of :

(A) Gluten enriched foods (B) Non-glutenous foods

(C) Phenol rich foods (D) Low gluten foods

531. Which salt has sweet taste?

(A) Sodium bicarbonate (B) Barium chloride

(C) Potassium iodide (D) Lead acetate

532. Which protein contains highest glutamic acid?

(A) Wheat gluten (B) Casein

(C) Corn gluten (D) Egg albumen

533. Waxy flavor is a characteristic of :
(A) Palmitic acid (B) Acetic Acid
(C) Lauric acid (D) Decanoic acid

534. Hesperidin is primarily isolated from:
(A) Lemon (B) Grapefruit
(C) Pummelo (D) Mandarin

535. Taste buds for bitterness are located at the ....of tongue.
(A) Sides (B) Tip
(C) Base (D) Both at tip and base

536. Which preservative is mainly used in baked products?
(A) Sorbic acid (B) SO_2
(C) Sodium benzoate (D) Sugar

537. Which grape variety is commercially used in baked products?
(A) White Riesling (B) Thompson Seedless
(C) Sundekhani (D) Pearl of Casaba

538. For apple cider making, the TSS of apple variety should be:
(A) 8-9 (B) 12-13
(C) 15-16 (D) 18-20

539. The DFFB, a maturity index used commercially for Delicious group of apples under Indian condition is:
(A) 88 ±4days (B) 126±4 days
(C) 150±4days (D) 180±4days

540. Strawberry fruit is derive from which plant tissue :
(A) Pericorp (B) Receplack
(C) Peddicle (D) Peduncle

541. Commercially, the chemical used to scrub ethylene from the fruit storage :
(A) Potassium oxalate
(B) Potassium sodium titrate

(C) Potassium dichromate

(D) Potassium permanganate

542. The hydrolysis of pectic substances during fruit ripening is brought about by the enzyme :

(A) Pectin methyl esterase (B) Polygalacturonase

(C) Cellulose (D) Both A and B

543. The red pigmentation of apple is due to :

(A) Cyaniding- 3- galactoside

(B) Ethyl2- methyl butyrate

(C) Pelargonidin-3-gluctoside

(D) Hesperidin

544. In climacteric fruit ethylene biosynthesis can be reduce by the application of :

(A) Silver nitrate

(B) Sodium thiosulphate

(C) 2-chloroethyl Phosphoric acid

(D) Aminoethoxyvinyl glycine

545. Which chilli has maximum SHU units in the world?

(A) Red Sabina

(B) Bhut Jaulokia

(C) Shimla Mirch

(D) Pusa Sadabahar

546. Among the following fruit, the highest lipid content is found in :

(A) Mango (B) Avocado

(C) Banana (D) Pineapple

547. As compared to fruit and acidic vegetable, the processing temperature for non acidic Vegetable is usually :

(A) Higher (B) Lower

(C) The same (D) Its make no different

548. For every increase of 152m in altitued, the normal processing time for fruit and vegetable has to be increased over the normal period of processing at sea-level by :

(A) 1 mint (B) 2 mint

(C) 4 mint (D) 8 mint

549. Which of the following in defect in products is caused by microorganisms.

(A) Flat sour (B) Hydrogen swell

(C) Springer (D) Flipper

550. Sugar syrup flavoured with artificial essences of fruits/ herbs is known :

(A) Sharbet (B) Squash

(C) Cordial (D) None of these

551. Pectinol, used for clarifying juice is :

(A) An enzyme (B) Pectin

(C) Protein (D) Starch

552. The pH of food which can be processed at 100°C should be below :

(A) 4.5 (B) 5.5

(C) 6.5 (D) 7.5

553. Which of the following are plotted on a psychometric chart?

(A) Temperature

(B) Vapour pressure of water

(C) Relative humidity

(D) None of these

554. A hermetically sealed container means, a container which is :

(A) Completely air tight

(B) Slightly air tight

(C) Allows for exchange of air

(D) None of these

555. Which of the Following is capable of delaying yellowing of leaves :

(A) Giberelic acid (B) IAA

(C) ABA (D) Cytokinins

556. Phytochrome is found in :

(A) Fungi (B) Algae

(C) Pteridophytes (D) Angiosperms

557. A flower with stamens, carpel, sepal, and petals is called :

(A) Incomplete Flower (B) Complete flower

(C) Hermaphrodite Flower (D) None of these

558. Which of the following concentration of sugar syrup is lowest to prevent the growth of microbes :

(A) 61 % (B) 66%

(C) 72 % (D) 76%

559. Drained weight of vegetable is :

(A) 45% (B) 50 %

(C) 55 % (D) 100%

560. Generally fruit are processed at :

(A) 76.8 ^{0}C (B) Super heated steam

(C) Pressure of 15 lbs/PSI (D) Boiling water

561. The invert sugar in finished jam should be :

(A) 30-40 % (B) 45-55 %

(C) 60-65 5 % (D) 65-68 %

562. Vegetable are subjected to drying after :

(A) Blanching (B) No blanching

(C) Sulphuring (D) Any other means

563. Grapes are generally dried in :

(A) Sun (B) Dehydrator

(C) Spray drying (D) Vacuum drier

564. Vegetable are packed in can processed by heat in first is :

(A) Retourt (B) Water bath

(C) Sun (D) Any of them

565. For pickle making one can use maximum :

(A) 40% salt (B) 30 % salt

(C) 25% salt (D) 35% salt

566. Jam is prepared with :

(A) Fruit pectin and pulp

(B) Fruit pulp alone

(C) Fruit pectin alone

(D) Pectin, pulp, seed and skin mix

567. β-oxidation occurs in

(A) Spherosomes (B) Peroxisomes

(C) Glycoxisome (D) Mitochondria

568. Water soluble sugar is :

(A) Glucose (B) Fructose

(C) Sucrose (D) All of this

569. Cane with mild positive pressure called as :

(A) Flipper (B) Springer

(C) Bursting of can (D) Leaker

570. Grape wine are two types :

(A) Strong and light

(B) Medium and light

(C) Dry and sweet

(D) Better and sweet

571. For the powdery or granular materials, the sampling techniques will be............

(A) Dilution (B) Heating

(C) Screening (D) Quartering

572. In .... sampling, coarse cutting blade and nut butter blade are used for grinding.

(A) Colour product (B) Frozen product

(C) Dried product (D) Wet product

573. During sampling of plant tissue of fruits and vegetables, the samples are dried.........

(A) Around 6% (B) Below 10%

(C) Above 10% (D) None of above

574. Fruits and fruit products for sugar analysis may be preserved for several days by adding excess of neutral ....

(A) Lead oxide (B) Lead acetate

(C) Sodium acetate (D) Potassium acetate

575. Water as a dispersing medium for colloids and as a solvent for the crystalloids present is known as ....

(A) Bond water (B) Capillary water

(C) Crystal water (D) Free water

576. For the determination of moisture the fruits should be dried at............under vacuum.

(A) 100-110°C (B) 60-70°C

(C) 100-110°K (D) 60-70°K

577. Spices should be dried at ....°C for 4 days at normal atmosphere.

(A) 110 (B) 70

(C) 66 (D) 55

578. ....is used as solvent in the determination of moisture of spices.

(A) Ethyl ether (B) Methyl ether

(C) Toluene (D) Water

579. ....is used for the determination of moisture content.

(A) Microwave oven (B) Spectrophotometer

(C) Centrifuge (D) Muffle furnace

580. Generally ash content is determined after the determination of ………… .

(A) Acidity (B) Moisture

(C) Sugar (D) pH

581. ………..is used for the determination of ash content.

(A) Microwave oven

(B) Spectrophotometer

(C) Centrifuge

(D) Muffle furnace

582. The …………..is a measures of sand and other siliceous matter present in sample.

(A) Acid-soluble Ash

(B) Water-soluble Ash

(C) Acid-insoluble Ash

(D) Water-insoluble Ash

583. ……… is ultimate criteria of the desirability of any food product to consumer.

(A) Quality (B) Availability

(C) Price (D) Stability

584. Generally for the determination of acidity in fruit juice, the …………….. % phenolphthalein is used as indicator during titration.

(A) 0.01 (B) 0.1

(C) 1.0 (D) 10

585. The ……….. quality is important to consumer as it satisfies his aesthetic and gustatory sense.

(A) Quantity (B) Sensory

(C) Packaging (D) Nutritional

586. Appearance of food can be judged by ………….. organ of senses.

(A) Eye (B) Nose

(C) Ear (D) Toung

587. Odour in food can be judged by .... organ of senses.

(A) Eye (B) Nose

(C) Ear (D) Toung

588. Texture of food can be judged by .... organ of senses.

(A) Eye (B) Nose

(C) Ear (D) Toung

589. Taste of food can be judged by .... organ of senses.

(A) Eye (B) Nose

(C) Ear (D) Toung

590. The trained panel usually constituted of .... members for sensory analysis.

(A) 5-10 (B) 25-30

(C) 100 (D) More than 100

591. The semi-trained panel usually constituted of............ members for sensory analysis.

(A) 5-10 (B) 25-30

(C) 100 (D) More than 100

592. The consumer panel usually constituted of .... members for sensory analysis..

(A) 5-10 (B) 25-30

(C) 100 (D) More than 100

593. The consumer trail usually done for ....

(A) Judging the acceptability of product before consumer trial

(B) Final product acceptability

(C) Research and development

(D) All of above

594. The semi-trained panel is also known as .... panel.

(A) Unskilled

(B) Consumer preference

(C) Discriminative and Communicative

(D) Smart but unskilled

595. The best time of sensory evaluation is ....

(A) 10.00 a.m. to 12.00 noon

(B) 3.00 p.m. to 5.00 p.m.

(C) 10.00 p.m. to 12.00 a.m. and 3.00 a.m. to 5.00 a.m.

(D) 10.00 a.m. to 12.00 noon and 3.00 p.m. to 5.00 p.m.

596. The best time of sensory evaluation is ....

(A) An hour after any normal meal

(B) An hour before any normal meal

(C) An hour before dinner or sleep

(D) An hour after dinner or sleep

597. The number of individual sample to be selected may be found from the expression ....

(A) $n=C(N)^{1/2}$ (B) $N=C(n)^{1/2}$

(C) $n=N(C)^{1/2}$ (D) $N= n(C)^{1/2}$

598. Chlorin compounds are the most popular santizers used in the food processing factors because of the following characteristics except : (IARI-2012)

(A) Activity against a wide variety of microorganisms

(B) Harmless residue left over

(C) Easily dissociation

(D) Stability in the presence of hard water salts

599. Water activity in a processed food usually denotes : (IARI-2012)

(A) Actual water content

(B) The availability of water for microbial growth

(C) The availability of water for enzyme activity, browning and oxidation

(D) Both (B) and (C)

600. Amongst all the hydrocarbons, ethylene is the most effective fruit ripening agent because it possesses :

(A) Short chain length and double bond

(B) Short chain length and triple bond

(C) Longer chain length and double bond

(D) Longer chain length and triple bond

Answer Key

1.	2.	3 .	4.	5.	6.	7.	8.	9.	10	11.	12.	13.	14.	15.
B	D	C	A	A	D	C	B	C	A	C	B	C	A	C
16.	17.	18.	19.	20.	21.	22.	23.	24.	25.	26.	27.	28.	29.	30.
B	C	B	D	A	C	B	C	B	A	B	B	D	A	C
31.	32.	33.	34.	35.	36.	37.	38.	39.	40.	41.	42.	43.	44.	45.
D	A	B	D	A	A	C	A	A	B	D	D	C	C	B
46.	47.	48.	49.	50.	51.	52.	53.	54.	55.	56.	57.	58.	59.	60.
B	A	A	B	D	A	A	B	C	B	A	A	D	A	D
61.	62.	63.	64.	65.	66.	67.	68.	69.	70.	71.	72.	73.	74.	75.
C	B	C	C	A	A	B	B	B	B	C	D	A	A	D
76.	77.	78.	79.	80.	81.	82.	83.	84.	85.	86.	87.	88.	89.	90.
B	A	C	B	B	D	C	B	A	B	D	D	A	A	A
91.	92.	93.	94.	95.	96.	97.	98.	99.	100.	101.	102.	103.	104.	105.
B	B	A	B	A	B	C	D	C	A	D	D	A	B	B
106.	107.	108.	109.	110.	111.	112.	113.	114.	115.	116.	117.	118.	119.	120.
C	C	D	C	B	A	A	B	A	B	A	B	C	D	B

121.	122.	123.	124.	125.	126.	127.	128.	129.	130.	131.	132.	133.	134.	135.
B	C	A	C	D	B	A	C	D	A	A	B	D	A	A
136.	137.	138.	139.	140.	141.	142.	143.	144.	145.	146.	147.	148.	149.	150.
D	B	D	D	B	B	A	C	C	B	C	A	D	A	C
151.	152.	153.	154.	155.	156.	157.	158.	159.	160.	161.	162.	163.	164.	165.
A	B	A	D	D	D	D	D	A	A	A	B	B	A	D
166.	167.	168.	169.	170.	171.	172.	173.	174.	175.	176.	177.	178.	179.	180.
A	D	B	D	A	B	C	A	A	B	C	C	C	B	D
181.	182.	183.	184.	185.	186.	187.	188.	189.	190.	191.	192.	193.	194.	195.
C	C	C	A	A	C	B	D	D	A	C	A	B	A	B
196.	197.	198.	199.	200.	201.	202.	203.	204.	205.	206.	207.	208.	209.	210.
B	C	B	D	B	A	D	B	D	C	A	D	B	B	B
211.	212.	213.	214.	215.	216.	217.	218.	219.	220.	221.	222.	223.	224.	225.
D	B	C	B	B	C	B	A	C	D	B	B	A	A	D
226.	227.	228.	229.	230.	231.	232.	233.	234.	235.	236.	237.	238.	239.	240.
C	D	B	D	C	B	C	B	A	D	D	A	C	C	D
241.	242.	243.	244.	245.	246.	247.	248.	249.	250.	251.	252.	253.	254.	255.
D	A	A	B	A	B	D	D	A	D	B	C	B	B	C

256.	257.	258.	259.	260.	261.	262.	263.	264.	265.	266.	267.	268.	269.	270.
C	C	D	C	A	D	C	D	C	D	C	D	A	B	C
271.	272.	273.	274.	275.	276.	277.	278.	279.	280.	281.	282.	283.	284.	285.
A	A	C	C	B	B	A	C	A	C	B	C	C	D	A
286.	287.	288.	289.	290.	291.	292.	293.	294.	295.	296.	297.	298.	299.	300.
C	C	A	D	D	C	D	D	B	B	B	B	B	D	C
301.	302.	303.	304.	305.	306.	307.	308.	309.	310.	311.	312.	313.	314.	315.
D	B	B	B	A	A	B	B	A	C	B	A	B	C	C
316.	317.	318.	319.	320.	321.	322.	323.	324.	325.	326.	327.	328.	329.	330.
D	C	B	A	B	A	D	C	D	A	D	A	B	B	C
331.	332.	333.	334.	335.	336.	337.	338.	339.	340.	341.	342.	343.	344.	345.
D	A	C	D	C	C	D	C	D	A	D	B	A	B	C
346.	347.	348.	349.	350.	351.	352.	353.	354.	355.	356.	357.	358.	359.	360.
C	B	C	D	C	A	B	A	B	C	A	D	A	A	C
361.	362.	363.	364.	365.	366.	367.	368.	369.	370.	371.	372.	373.	374.	375.
B	C	B	D	B	A	B	B	D	B	B	A	A	C	B
376.	377.	378.	379.	380.	381.	382.	383.	384.	385.	386.	387.	388.	389.	390.
C	C	C	C	B	B	B	C	D	A	D	A	D	C	B

391.	392.	393.	394.	395.	396.	397.	398.	399.	400.	401.	402.	403.	404.	405.
A	A	C	A	D	B	A	A	B	D	A	B	A	C	D
406.	407.	408.	409.	410.	411.	412.	413.	414.	415.	416.	417.	418.	419.	420.
D	D	C	A	B	B	D	C	A	A	B	D	D	C	D
421.	422.	423.	424.	425.	426.	427.	428.	429.	430.	431.	432.	433.	434.	435.
A	B	D	C	A	B	D	C	B	B	B	D	D	C	C
436.	437.	438.	439.	440.	441.	442.	443.	444.	445.	446.	447.	448.	449.	450.
C	D	D	D	C	B	C	C	A	C	C	D	B	B	D
451.	452.	453.	454.	455.	456.	457.	458.	459.	460.	461.	462.	463.	464.	465.
D	A	C	D	B	D	C	B	B	C	C	C	C	D	B
466.	467.	468.	469.	470.	471.	472.	473.	474.	475.	476.	477.	478.	479.	480.
B	B	C	C	B	D	D	A	B	B	C	D	A	C	A
481.	482.	483.	484.	485.	486.	487.	488.	489.	490.	491.	492.	493.	494.	495.
D	D	B	C	A	D	C	B	A	B	A	C	A	D	C
496.	497.	498.	499.	500.	501.	502.	503.	504.	505.	506.	507.	508.	509.	510.
C	D	A	B	C	B	D	A	B	D	A	B	C	B	A

511.	512.	513.	514.	515.	516.	517.	518.	519.	520.	521.	522.	523.	524.	525.
D	C	C	D	D	C	B	D	C	B	C	C	A	B	C
526.	527.	528.	529.	230.	531.	532.	533.	534.	535.	536.	537.	538.	539.	540.
C	B	A	B	A	A	A	A	A	C	A	B	D	B	B
541.	542.	543.	544.	545.	546.	547.	548.	549.	550.	551.	552.	553.	554.	555.
D	D	A	D	B	B	A	B	A	A	A	A	D	A	D
556.	557.	558.	559.	560.	561.	562.	563.	564.	565.	566.	567.	568.	569.	570.
D	B	C	C	D	A	A	A	A	C	A	A	C	A	C
571.	572.	573.	574.	575.	576.	577.	578.	579.	580.	581.	582.	583.	584.	585.
D	C	A	B	D	B	D	C	A	B	D	C	A	C	B
586.	587.	588.	589.	590.	591.	592.	593.	594.	595.	596.	597.	598.	599.	600.
A	B	C	D	A	B	D	B	C	D	A	A	B	D	A

❑❑❑

29

Appendices

Appendix-I

Fruits: Common name, Botanical name, Family, Chromosome No., Origin, Type of fruit and Edible Portion

S.No.	Common name	Botanical name	Family	2n No.	Origin	Type offruit	EdiblePortion
1.	Kiwi fruit	*Actinidia deliciosa*	Actinidaceae	58	China	Berry	Epicarp & Mesocarp
2.	Chinese gooseberry	*Actinidia chinensis*	Actinidaceae	58, 174	China	Berry	Epicarp & Mesocarp
3.	Arecanut	*Areca catechu*	Arecaceae	32	Malayan Archipelago	Single seeded berry	Seed

Contd.

4.	Coconut	*Cocus nucifera*	Arecaceae	32	South East Asia (Malaysia, Indonesia)	One seeded Drupe	Endosperm
5.	Date palm	*Phoenix dactylifera*	Arecaceae	36	West Asia (Iraq)	Drupe	Pericarp
6.	Oil palm	*Ealias guinensis*	Arecaceae	32	Africa	Berry	Mesocarp
7.	Palmyra palm	*Borasus flaballifer*	Arecaceae	32	Tropical Africa	3 seeded drupe	Mesocarp
8.	Mango	*Mangifera indica*	Anacardiaceae	40	Indo-Burma	Drupe	Mesocarp
9.	Amada	*Spondias Mangifera*	Anacardiaceae			Drupe	Mesocarp
10.	Cashew nut	*Anacardium occidentale*	Anacardiaceae	42	Brazil	Nut	Cotyledon & fleshy peduncle
11.	Pistachio nut	*Pistacia vera*	Anacardiaceae	30	West Asia	Nut	Cotyledon
12.	Atemoya	*Annona atemoya*	Annonaceae	14	Man made hybrid	Etaerio of berries	Pericarp of individual fruit
13	Custard apple	*Annona squamosa*	Annonaceae	14	Tropical America	Etaerio of berries	Pericarp of individual fruit
14.	Soursop	*Annona muricata*	Annonaceae	14	–	Etaerio of berries	Pericarp of individual fruit
15.	Pond apple	*Annona glabra*	Annonaceae	14	–	Etaerio of berries	Pericarp of individual fruit
16.	Bullock's heart	*Annona reticulata*	Annonaceae	14	India	Etaerio of berries	Pericarp of individual fruit
17.	Cherimoya	*Annona cherimola*	Annonaceae	14	–	Etaerio of berries	Pericarp of individual fruit

Contd.

18.	Ilma	*Annona diversifolia*	Annonaceae	14	–	Etaerio of berries	Pericarp of individual fruit
19.	Karonda	*Carissa carandas*	Apocynaceae	22	India Java	Berry	Epicarp & Mesocarp
20.	Natal palm	*Carissa grandiflora*	Apocynaceae	–	–	–	–
21.	Pineapple	*Annanas comosus*	Bromeliaceae	50	Brazil	Sorosis	Bracts & Periantus
22.	Papaya	*Carica papaya*	Caricaceae	18	Tropical America	Berry	Mesocarp
23.	Hazelnut (Bhatia Badam)	*Corylus avellana*	Corylaceae	–	–	Nut	–
24.	Persimmon	*Diospyros kaki*	Ebenaceae	90	China	Berry	Epicarp & Mesocarp
25.	Aonla	*Emblica officinalis*	Euphorbiaceae	28	Tropical Asia (Indo-China)	Berry	Mesocarp
26.	Chestnut	*Castanea sativa*	Fagaceae	24		Nut	
27.	Mangosteen	*Garcinia mangostana*	Guttiferae	24		Berry	
28.	Pecan nut	*Carya illinoensis*	Juglandaceae	36	Southern USA	Nut	Seed or cotyledon
29.	Walnut	*Juglans regia*	Juglandaceae	32	Iran	Nut	Indocorp and Mesocorp
30.	Avocado	*Persia americana*	Lauraceae	24	Mexico & West Indies	Berry	Pericarp
31.	Tamarinds	*Tamarindus indica*	Leguminosae	24	Tropical Africa-India	Pod	Pulp/Mesocarp
32.	Barbados cherry	*Malpighia glabra*	Malpighiaceae			Berry	
33.	Jack fruit	*Artocarpus heterophyllus*	Moraceae	56	India	Sorosis	Bracts, Seed or Periantus
34.	Monkey jack	*A. lakoocha*	Moraceae			Sorosis	Bracts, Seed or Periantus

Contd.

35.	Bread fruit	*A. altilis*	Moraceae			Sorosis	
36.	Fig	*Ficus carica*	Moraceae	26	West Asia	Syconus	Fleshy receptacle
37.	Mulberry (white)	*Morus alba*	Moraceae	308	China	Syncarpous/ aggregate/ sorosis	Mesocarp
38.	Mulberry (black)	*M. nigra*	Moraceae	308	China	Syncarpous/ aggregate/ sorosis	Mesocarp
39.	Mulberry (red)	*M. rubra*	Moraceae	308	China	Syncarpous/ aggregate/ sorosis	Mesocarp
40.	Banana	*Musa paradidaca*	Musaceae	22, 33,44	South East Asia	Berry	Mesocarp & endocarp
41.	Guava	*Psidium guajava*	Myrtaceae	22	West Indies to Peru	Berry	Thalamus & pericarp
42.	Strawberry guava	*P. cattelianum*	Myrtaceae			Berry	Thalamus & pericarp
43.	Jamun	*Syzygium cumunii*	Myrtaceae	40	East India-Malaya	Drupe	Epicarp & Mesocarp
44.	Rose apple	*S. jambos*	Myrtaceae			Drupe	Epicarp & Mesocarp
45.	Surinam cherry	*Eugenia uniflora*	Myrtaceae				
46.	Olive	*Olea europaea*	Oleaceae	46	Mediterra -nean region	Drupe	Epicarp & Mesocarp
47.	Carambola	*Averrhoa carambola*	Oxalidaceae	24	India-China	Berry	Pericarp
48.	Passion fruit	*Passiflora edulis*	Passifloraceae	18	Brazil (Tropical America)	Berry	Mesocarp

Contd.

49.	Pomegranate	*Punica granatum*	Punicaceae	18	Iran	Balausta	Juicy seed coat (Aril)
50.	Macadamia nut	*Macadamia ternifolia*	Proteaceae	48	Australia	Nut	
51.	Ber (cultivated)	*Zizyphus mauritiana*	Rhamnaceae	48(4x)	India	Drupe	Pericarp
52.	Ber (Chinese)	*Z. jujuba*	Rhamnaceae	48(4x)	China	Drupe	Pericarp
53.	Apple	*Malus pumila*	Rosaceae	34	Asia Minor to Western Himalayas	Pome	Thalamus
54.	Almond	*Prunus communis*	Rosaceae	16	Persia-Afghanistan	Drupe	Kernel
55.	Apricot	*Prunus armeniaca*	Rosaceae	16	China	Drupe	Mesocarp & endocarp
56.	Black berry	*Rubus fruticosus*	Rosaceae				
57.	Crab apple	*Malus baccata*	Rosaceae			Pome	Thalamus
58.	Loquat	*Eriobotrya japonica*	Rosaceae	34	China	Pome	Fleshy Thalamus
59.	Pear	*Pyrus communis*	Rosaceae	34	Mountain region of Western China	Pome	Fleshy Thalamus
60.	Peach	*Prunus persica*	Rosaceae	16	China	Drupe	Epicarp & Mesocarp
61.	Plum (Japanese)	*Prunus domestica*	Rosaceae	48(6x)	China	Drupe	Epicarp & Mesocarp
62.	Plum (European)	*Prunus salicina*	Rosaceae	16	Europe	Drupe	Epicarp & Mesocarp
63.	Quince	*Cydonia oblonga*	Rosaceae	34	South East Europe to Asia Minor	Drupe	Fleshy Thalamus
64.	Strawberry	*Fragaria ananasa*	Rosaceae	56	Mane made hybrid	Etaerio of achenes	Succulent thalamus

Contd.

65.	Sweet cherry	*Prunus avium*	Rosaceae	16	Europe West Asia	Drupe	Epicarp & Mesocarp
66.	Sour cherry	*Prunus cerasus*	Rosaceae	16	Europe West Asia	Drupe	Epicarp & Mesocarp
67.	Coffee	*Coffea arabica*	Rubiaceae	44	Ethiopia	Drupe	Seed
68.		*Coffea robusta*	Rubiaceae	22	Ethiopia	Drupe	Seed
69.	Bael	*Aegle marmelos*	Rutaceae	36(4x)	India	Amphisarca	Succulent placenta
70.	Citron	*Citrus medica*	Rutaceae	16		Hesperidium	
71.	Grapefruit	*Citrus paradise*	Rutaceae	16	West Indies	Hesperidium	Juicy placental hairs
72.	Hazara	*Citrus myrtifolia*	Rutaceae	16	India	Hesperidium	Juicy placental hairs
73.	Karna khatta	*Citrus karna*	Rutaceae	16	India	Hesperidium	Juicy placental hairs
74.	Jambhiri	*Citrus jambhiri*	Rutaceae	16	Indai	Hesperidium	Juicy placental hairs
75.	Kumquat (oval)	*Fortunella margarita*	Rutaceae	16	–	Hesperidium	Juicy placental hairs
76.	Kumquat (round)	*F. japonica*	Rutaceae	16	–	Hesperidium	Juicy placental hairs
77.	Lemon	*Citrus limon*	Rutaceae	16	China	Hesperidium	Juicy placental hairs
78.	Lime	*Citrus aurantifolia*	Rutaceae	16	India	Hesperidium	Juicy placental hairs
79.	Mandarin	*Citrus reticulate*	Rutaceae	16	China	Hesperidium	Juicy placental hairs
80.	Rangpur lime	*Citrus limonia*	Rutaceae	16	India	Hesperidium	Juicy placental hairs
81.	Sweet lime	*Citrus limettoides*	Rutaceae	16	India	Hesperidium	Juicy placental hairs
82.	Sour orange	*Citrus aurantium*	Rutaceae	16	China	Hesperidium	Juicy placental hairs
83.	Sweet orange	*Citrus sinensis*	Rutaceae	16	China	Hesperidium	Juicy placental hairs
84.	Pummelo	*Citrus grandis*	Rutaceae	16	South East Asia	Hesperidium	Juicy placental hairs

Contd.

85.	Wood apple	*Citrus acidissima*	Rutaceae	16	India	Amphisiaraca	Succulent placentae
86.	Rambutan	*Nephelium lappaceum*	Sapindaceae	22	Malaya	Berry	Aril
87.	Litchi	*Litchi chinensis*	Sapindaceae	30	China	Nut	Fleshy aril
88.	Sapota	*Achrus zapota*	Sapotaceae	26	Mexico (Tropical America)	Berry	Mesocarp
89.	Cape gooseberry	*Physalis peruviana*	Solanaceae	24, 48		Berry	Mesocarp & endocarp
90.	Cocoa	*Theoborma cocoa*	Sterculiaceae	20	Amazon valley of South America	Drupe(Pod)	Seed
91.	Tea	*Camelia sinensis*	Theaceae/ Camaliaceae	30	China		Leaves
92.	Phalsa	*Grewia subinequalis*	Tilliaceae	36(4x)	India	Drupe	Epicarp & Mesocarp
93.	Grape (European)	*Vitis vinifera*	Vitaceae	38	Black sea to Caspian sea	Berry	Pericarp & Placentae
94.	Grape (American)	*Vitis labrusca*	Vitaceae	38	Black sea to Caspian sea	Berry	Pericarp & Placentae
95.	Grape (Muscadine)	*Vitis rotundifolia*	Vitaceae	38	Black sea to Caspian sea	Berry	Pericarp & Placentae

Appendix-II

Vegetables: Common name, Botanical name, Family, Chromosome No., Origin, Type of fruit and Edible Portion

S.N.	Common name	Botanical name	Family	Chromosome No.(2n)	Origin	Edible Portion
			Monocotyledonous			
1.	Chive	*Allium schenoprasum*	Alliaceae	16,24,32		Hollow green leaves and bulbs
2.	Chinese Chive	*Allium tuberosum*	Alliaceae	4x=32	China	Green leaves
3.	Garlic	*Allium sativum*	Alliaceae	16	Central Asia	Clove
4.	Great Headed Garlic	*Allium ampeloprasum*	Alliaceae	4x=32,6x=48		Clove and leaves
5.	Kurrat	*Allium kurrat*	Alliaceae	4x=32		Green leaves
6.	Leek	*Allium porrum*	Alliaceae	4x=32	Central Asia	Green leaves, leaf base and root
7.	Onion	*Allium cepa*	Alliaceae	16	Central Asia	Bulb
8.	Potato/multiplier Onion	*Allium cepa var. aggregatum*	Alliaceae	16	–	Bulb
9.	Rakkyo	*Allium chinense*	Alliaceae	16,48	–	Bulb
10.	Shallot	*Allium cepa var. ascalonicum*	Alliaceae	16	–	Bulb
11.	Tree/Proliferum/ Egyptian tree Onion	*Allium cepa var. viviparium/proliferum*	Alliaceae	4x=32	–	Bulb
12.	Welsh/Japanese bunching onion	*Allium fistulosum*	Alliaceae	16	–	Leaf top and its base

Contd.

13.	Elephant Foot Yam	*Amorphophyllus campanulatus*	Araceae	26	–	Corms
14.	Colocasia	*Colocasia esculenta*	Araceae	28	Srilanka	Leaves and tuberous corms
15.	Colocasia	*Colocasia esculenta var. globulifera*	Araceae	28	Srilanka	Leaves and tuberous corms
16.	Tannia	*Xanthosoma sagittifolium*	Araceae	26	–	Corms
17	Amorphophallus	*Amorphophallus companulatus*	Araceae	26	–	corn
18.	Greater Yam	*Dioscorea alata*	Dioscoraceae	40	South East Asia	Tubers
19.	Lesser Yam	*Dioscorea esculenta*	Dioscoraceae	40	Indo-Burma	Tubers
20.	White Yam	*Dioscorea rotundata*	Dioscoraceae	40	Indo-Burma	Tubers
21.	Sweet Corn	*Zea mays*	Graminae	20		Kernel
22.	Asparagus	*Asparagus officinalis*	Lilliaceae	20	Europe, Asia	Spwears
			Dicotyledonous			
23.	Amaranthus (Badi Chulai)	*Amaranthus tricolor*	Amaranthaceae	32,34	India (South and South East Asia)	Leaves, Stem, Seeds
24.	Amaranthus (Choti Chulai)	*Amaranthus blitum*	Amaranthaceae	32	India (South and South East Asia)	Leaves, Stem, Seeds
25.B	Beet Root	*Beta vulgaris var. rubra*	Chenopodiaceae	18	Mediterranean region	Swollen root
26.	Palak	*Beta vulgaris va. Bengalensis*	Chenopodiaceae	18	Indo-Chinese region	Leaves
27.	Spinach (Vilayati Palak)	*Spinacea oleracea*	Chenopodiaceae	12	Iran (South West Asia)	Leaves

Contd.

28.	Bathua (Pigweed)	*Chenopodium album*	Chenopodiaceae			Leaves
29.	Lettuce (Head type)	*Lactuca sativa var. capitata*	Compositae	18	Mediterranean region	Leaves and Stem
30.	Lettuce (Leafy/ open type)	*Lactuca sativa var. crispa*	Compositae	18	Mediterranean region	Leaves and Stem
31.	Globe artichoke	*Cynara artichoke*	Compositae	34	Mediterranean region	Bud scale and flower head
32.	Jerusalem Artichoke /Girasole	*Helianthus tuberosus*	Compositae	102	North America (USA)	Underground tubers
33.	Endive	*Cichorium endiva*	Compositae	18,36	Egypt or India	Leaves
34.	Chicory		Compositae	18		Leaves
35.	Sweet Potato	*Ipomea batatas*	Convolvulaceae	6x=90	Mexico and Central America	Roots
36.	Cabbage	*Brassica oleracea var. capitata*	Cruciferare/ Brassicaceae	18	Mediterranean region	Head
37.	Cauliflower	*Brassica oleracea var. botrytis*	Brassicaceae	18	Mediterranean region	Cured
38.	Brussel's sprout	*Brassica oleracea var. gemmifera*	Brassicaceae	18	Mediterranean region	Buds
39.	Sprouting Broccoli	*Brassica oleracea var. italica*	Brassicaceae	18	Mediterranean region	Flower bud
40.	Knol-Khol	*Brassica oleracea var. gongylodes or B. caulorapa*	Brassicaceae	18	Mediterranean region	Swollen stem
41.	Kale	*Brassica oleracea var. acephgala*	Brassicaceae	18	Mediterranean region	Leaves and shoot

Contd.

42.	Kitchen Kale	*Brassica oleracea var. fimbriata*	Brassicaceae	18	–	Leaves
43.	Marrow stem kale	*Brassica oleracea var. medullosa*	Brassicaceae	18	–	Leaves
44.	Chinese Cabbage	*Brassica chinensis*	Brassicaceae	20	China	Leafy Head
45.	Turnip	*Brassica repa var. rapifera*	Brassicaceae	20	China, India	
46.	Radish	*Raphanus sativus*	Brassicaceae	18	Europe	Root and leaves
47.	Rutabga/Swedish turnip	*Brassica napus var. napobrassica*	Brassicaceae	38	–	Roots
48.	Leaf mustard	*Brassica juncea var. cunefolia*	Brassicaceae	36	Tender leaves	
49.	Water Cress	*Nasturtium officinale, N. microphyllum*	Brassicaceae	32	–	Leaves and young top end
50.	Horseradish	*Armoracia rusticana*	Brassicaceae	18	Eastern Europe	Leaves
51.	Cucumber	*Cucumis sativus*	Cucurbitaceae	14	India	Unripe or tender fruit
52.	Musk Melon	*Cucumis melo*	Cucurbitaceae	24	Tropical Africa	Ripe Fruit
53.	Snapmelon/ Phoot	*Cucumis melo var. momordica*	Cucurbitaceae	24	Tropical Africa	Ripe Fruit
54.	Long melon (kakri)	*Cucumis melo var. utillsiumus*	Cucurbitaceae	24	Tropical Africa	Unripe Fruit
55.	Gherkin	*Cucumis anguria*	Cucurbitaceae	24	Tropical Africa	Fruit
56.	Water melon	*Citrullus lanatus*	Cucurbitaceae	22	Tropical Africa	Ripe Fruit
57.	Round melon	*Citrullus lanatus var. fistulosus*	Cucurbitaceae	24	Tropical Africa	Unripe Fruit

Contd.

58.	Pumpkin	*Cucurbita moschata*	Cucurbitaceae	40	Mexico, Peru	Green and Ripe Fruit
59.	Summer sequash	*Cucurbita pepo*	Cucurbitaceae	40	Mexico	Fruit
60.	Winter squash	*Cucurbita maxima*	Cucurbitaceae	40	Mexico	Fruit
61.	Bottle gourd	*Legenaria siceraria*	Cucurbitaceae	22	South Africa / India	Fruit
62.	Bitter Gourd	*Momordica charantia*	Cucurbitaceae	22	Indo-Burma	Fruit
63.	Kakrol	*Momordica cochinchinenesis*	Cucurbitaceae	14	Asia and Africa	Unripe fruit
64.	Kartoli	*Momordica dioca*	Cucurbitaceae	28	India	Unripe fruit
65.	Ridge gourd	*Luffa acutangula*	Cucurbitaceae	26	Asia	Fruit
66.	Sponge gourd	*Luffa cylindrica*	Cucurbitaceae	26	Assam	Fruit
67.	Pointed gourd	*Trichosanthus dioca*	Cucurbitaceae	22	India	Fruit
68.	Snake Gourd	*Trichosanthus anguina*	Cucurbitaceae	22	India	Fruit
69.	Wax (Ash) gourd	*Benincasa hispida*	Cucurbitaceae	24	Java, Japan	Fruit
70.	Ivy Gourd	*Coccinia indica*	Cucurbitaceae	24	India	Fruit
71.	Chow-chow	*Sechium edule*	Cucurbitaceae	28	Mexico	Fruit
72.	Spine Gourd	*Momordia cochinchinesis*	Cucurbitaceae	28	–	Fruit
73.	Chayote	*Sechium edule*	Cucurbitaceae	28	South Mexico	Unripe fruit
74.	Cassava	*Manihot esculanta*	Euphorbiaceae	36	Brazil	Tuberous root
75.	Chekurmanis	*Sauropus androgynous*	Euphorbiaceae		Indo-Burma	Green leaves
76.	Chinese Potato	*Coleus pervriflorus*	Labitae		Africa	Tuber
77.	Pea	*Pisum sativum*	Leguminoceae / Fabaceae	14	Central Asia	Tender seed

Contd.

78.	French Bean	*Phaseolus vulgaris*	Leguminoceae	22	Mexico	Pod, Seed
79.	Asparagus Bean	*Vigna sinensis var. sesquipedalis*	Leguminoceae	22	Central Africa and India	Green pod, Seed
80.	Cluster Bean	*Cyampopsis tetragonolobus*	Leguminoceae	14	India	Pod, Seed
81.	Winged Bean	*Psophocarpus tetragonolobus*	Leguminoceae	18	Africa	Pod, Seed
82.	Lima Bean	*Phaseolus lunatus*	Leguminoceae		Gutmala	Pod
83.	Dolichos Bean	*Lablab purpureus*	Leguminoceae	22	India	Pod
84.	Cow Pea	*Vigna unguiculata*	Leguminoceae	22	Africa	Pod
85.	Agathi	*Sesbania grandiflora*	Leguminoceae	24	–	Flower
86.	Fenugreek	*Trigonella foenu graceum*	Leguminoceae	16	–	Leaves
87.	Winged Bean	*Psophocarpus tetragonolobus*	Leguminoceae	18	Africa	Pod
88.	Broad/Faba Bean	*Vicia faba*	Leguminoceae	12,14	Europe and Asia	Green pod, Seed
89.	Soybean	*Glycin max*	Leguminoceae	40	–	Dry seeds
90.	Okra	*Abelmoschus esculantus*	Malvaceae	130	Africa	Fruit
91.	Drumstick	*Moringa oleifera*	Moringaceae	28	Africa, India	Green pod
92.	Potato	*Solanum tuberosum*	Solanaceae	48	South America	Stem tuber
93.	Tomato	*Lycopersicon esculentum*	Solanaceae	24	South America	Fruit
94.	Brinjal	*Solanum melongena*	Solanaceae	24	Indo-Burma	Fruit
95.	Sweet pepper	*Capsicum*	Solanaceae	24	Mexico	Fruit

Contd.

96.	Chilli	*Capsicum*	Solanaceae	24	Mexico	Fruit
97.	Carrot	*Daucus carota*	Umbeliferae	18	Afganisthan	Root
98.	Corinder	*Coriandrum sativum*	Umbeliferae	22		Leaves
99.	Celery	*Apium graveolens*	Umbeliferae	22	Mediterranean region	Leaves
100.	Celeriac	*Apium graveolens var. rapaceum*	Umbeliferae	22	Mediterranean region	Bulbous roots
101.	Parsley	*Petroselinum crispum*	Umbeliferae	22	Mediterranean region	Leaves and seed
102.	Parsnip	*Pastinaca sativa*	Umbeliferae	22		Leaf fleshy tap root
103.	Parsley	–	–	–	–	–
104.	Bay leaf	*Paurus nobilis*	Lauraceae	–	Mediterranean region	Leaves
105.	New Zealand	*Tetragonia tetragonioides*	Aizoaceae		New Zealand	Tender leaves and tops
106.	Indian spinach	*Basella rubra*	Basillaceae	24	India	Fleshy stem and root
107.	Amorphophallus	*Amorphophallus companulatus*				

Appendix-III

Spices: Common name, Botanical name, Family, Chromosome No., Origin and Edible Portion.

Sr.no.	English Common Name	Family Name	Botanical Name	Chromosome No (2n)	Part use asspices
1	Cardamom (Small)	*Elettaria cardamomum Maton*	Zingiberaceae	48	Fruit, Seed
	Cardamom (Large)	*Amomum subulatum Roxb.*	Zingiberaceae	48	Fruit, Seed
2	Pepper	*Piper nigrum L.*	Piperaceae	128	Fruit
3	Chilli	*Capsicum*	Solanacea	24	Fruit
	Capsicum	*Capsicum frutescens L.*	Solanaceae	24	Fruit
	Chilli	*Capsicum annuum L.*	Solanaceae	24	Fruit
4	Ginger	*Zingiber officinale Rosc.*	Zingiberaceae	22	Rhizome
5	Turmeric	*Curcuma longa L.*	Zingiberaceae	62	Rhizome
6	Coriander	*Coriandrum sativum L.*	Apiaceae	22	Leaf & Fruit
7	Cumin	*Cuminum cyminum L.*	Apiaceae	14	Fruit
8	Fennel	*Foeniculum vulgare Mill.*	Apiaceae	22	Fruit
9	Fenugreek	*Trigonella foenum-graecum L.*	Fabaceae	14	Seed
10	Celery	*Apium graveolens L.*	Apiaceae	22	Leaf, Fruit & Stem
11	Aniseed	*Pimpinella anisum L.*	Apiaceae	40	Fruit
12	Ajowan	*Trachyspermum ammi L.*	Apiaceae	18	Fruit
13	Caraway	*Carum carvi L.*	Apiaceae	20	Fruit
14	Dill	*Anethum graveolens L.*	Apiaceae	22	Fruit

Contd.

15	Cinnamon	*Cinnamomum zeylanicum Breyn*	Lauraceae	24	Bark
16	Cassia	*Cinnamomum cassia.Blume*	Lauraceae	24	Bark
17	Garlic	*Allium sativum L.*	Alliaceae	16	Bulb
18	Curry leaf	*Murraya koenigii(L) Sprengel*	Rutaceae	18	Leaf
19	Kokam	*Garcinia indica Choisy*	Clusiaceae	48	Rind
20	Mint	*Mentha piperita L.*	Lamiaceae		Leaf
21	Mustard	*Brassica juncea L.Czern*	Brassicaceae	4x=36	Seed
22	Parsley	*Petroselinum crispum Mill.*	Apiaceae	22	Leaf
24	Saffron	*Crocus sativus L.*	Iridaceae	2n=3x=24	Stigma
25	Vanilla	*Vanilla planifolia Andr.*	Orchidaceae	20	Pod
26	Tejpat	*Cinnamomum tamala (Buch Ham) Nees & Eberum*	Lauraceae	20,32	Bark & Leaf
27	Pepper Long	*Piper longum L.*	Piperaceae	24,48,52	Fruit
28	Star Anise	*Illicium verum Hook.*	Illiciaceae		Fruit
29	Sweet flag	*Acorus calamus L.*	Araceae	18,24,36,44	Rhizome
30	Greater Galanga	*Alpinia galanga Willd.*	Zingiberaceae	48	Rhizome
31	Horse Radish	*Armoracia rusticana Gaertn.*	Brassicaceae	14	Root
32	Caper	*Capparis spinosa L.*	Capparidaceae	18	Flower buds
33	Clove	*Syzygium aromaticum(L) Merr.& Perry*	Myrtaceae	44	UnopenedFlower bud
34	Asafoetida	*Ferula asafoetida L*	Apiaceae		Oleogum resin from rhizome and thickened root

Contd.

35	Camboge	*Garcinia cambogia (Gaertn).Desr*	Clusiaceae		Rind
36	Hyssop	*Hyssopus officinalis L.*	Lamiaceae	24	Leaf
37	Juniper berry	*Juniperus communis L.*	Cupressaceae	22	Berry
38	Bay Leaf	*Laurus nobilis L.*	Lauraceae	36,42	Leaf
39	Lovage	*Levisticum officinale Koth.*	Apiaceae		Leaf & Stem
40	Marjoram	*Marjorana hortensis Moench.*	Lamiaceae		Leaf
41	Nutmeg	*Myristica fragrans Houtt.*	Myristicaceae	42	Seed
42	Mace	*Myristica fragrans Houtt.*	Myristicaceae	42	Aril
43	Basil	*Ocimum basilicum L.*	Lamiaceae	48	Leaf
44	Poppy seed	*Papaver somniferum L.*	Papaveraceae	22	Seed
45	Allspice	*Pimenta dioica (L) Merr.*	Myrtaceae	22	Fruit & Leaf
46	Rosemary	*Rosmarinus officinalis L.*	Lamiaceae	24	Leaf
47	Sage	*Salvia officinalis L.*	Lamiaceae	14	Leaf
48	Savory	*Satureja hortensis L.*	Lamiaceae		Leaf
49	Thyme	*Thymus vulgaris L.*	Lamiaceae		Leaf
50	Oregano	*Origanum vulgare L.*	Lamiaceae		Leaf
51	Tarragon	*Artemisia dracunculus L.*	Asteraceae	18	Leaf
52	Tamarind	*Tamarindus indica L.*	Caesalpiniaceae	24	Fruit

Appendix-IV

Energy and Nutritive Value of Fruits

S.NO.	Fruit	Energy (Kcals)	Moisture (g)	Protein (g)	Fat (g)	Mineral (g)	Fibre (g)	Carbohy -drates (g)	Calcium (mg)	Phosphorus (mg)	Iron (mg)
1.	Anola	58	82	0	0	0	3	14	50	20	1
2.	Apple	59	85	0	0	0	1	13	10	14	1
3.	Apricot fresh	53	85	1	0	1	1	12	20	25	2
4.	Apricot dry	306	19	2	1	3	2	73	110	70	5
5.	Avocado	215	74	2	23	1	0	1	10	80	1
6.	Bael	137	61	2	0	2	3	32	85	50	1
7.	Banana ripe	116	70	1	0	1	0	27	17	36	0
8.	Bread fruit	71	79	1	0	1	2	16	40	30	0
9.	Bullock heart	70	77	1	0	1	3	16	10	10	1
10.	Cape gooseberry	53	83	2	0	1	3	11	10	67	2
11.	Cashew fruit	51	86	0	0	0	1	12	10	10	0
12.	Cherries, red	64	83	1	0	1	0	14	24	25	0
13.	Cherries, black	316	18	3	0	2	1	75	130	110	8
14.	Date fresh	144	59	1	0	1	4	34	22	38	1
15.	Dates dried	617	15	2	0	2	4	76	120	50	7
16.	Grape pale green	71	79	0	0	1	3	16	20	30	0

Contd.

17.	Grape fruit seed less	45	88	1	0	0	0	10	30	30	0
18.	Grape fruit triumph	32	92	1	0	0	0	7	20	20	0
19.	Guava, country	51	82	1	0	1	5	11	10	28	0
20.	Guava, hill	38	85	0	0	1	5	9	50	20	1
21.	Jack fruit	88	76	2	0	1	1	20	20	41	0
22.	Lemon	57	85	1	1	0	2	11	70	10	0
23.	Lemon sweet	35	90	1	0	0	0	7	30	20	1
24.	Fig	37	88	1	0	1	2	8	80	30	1
25.	Litchi	61	84	1	0	0	0	14	10	35	1
26.	Lime	59	85	1	1	1	1	11	90	20	0
27.	Loquat	43	88	1	0	0	1	10	30	20	1
28.	Mango	74	81	1	0	0	1	17	14	16	1
29.	Mangosteen	60	85	1	0	0	0	14	10	20	0
30.	Musk melon	17	95	0	0	0	0	3	32	14	1
31.	Mulberry	49	86	1	0	1	1	10	70	30	2
32.	Orange	48	88	1	0	0	0	11	26	20	0
33.	Palmyra palm	87	77	1	0	1	0	21	9	33	0
34.	Papaya	32	91	1	0	0	1	7	17	13	0
35.	Passion fruit	54	76	1	0	1	10	12	10	60	2
36.	Peach	50	86	1	0	1	1	10	15	41	2
37.	Pears	52	86	1	0	0	1	12	8	15	0

Contd.

38.	Persimmon	76	80	1	0	0	1	18	15	10	0
39.	Phalsa	72	81	1	1	1	1	15	129	39	3
40.	Pineapple	46	88	0	0	0	0	11	20	9	2
41.	Plum	52	87	1	0	0	0	11	10	12	1
42.	Pomegranate	65	78	2	0	1	5	14	10	70	2
43.	Prunes	56	8	0	0	1	0	13	10	18	0
44.	Pummelo	44	88	1	0	0	1	10	30	30	0
45.	Quince	50	86	0	0	0	2	12	20	10	0
46.	Raisins	308	20	2	0	2	1	75	87	80	8
47.	Raspberry	56	85	1	1	1	1	12	40	110	2
48.	Rose apple	39	89	1	0	0	1	8	10	30	0
49.	Sapota	98	74	1	1	0	3	21	28	27	1
50.	Strawberry	44	88	1	0	0	1	10	30	30	2
51.	Tomato ripe	20	94	1	0	0	1	4	48	20	1
52.	Wood Apple	134	64	7	4	2	5	18	130	110	0

Appendix-V

Energy and Nutritive Value of Vegetable

S.NO.	Vegetable	Energy (Kcals)	Moisture (g)	Protein (g)	Fat (g)	Mineral (g)	Fibre (g)	Carbohy -drates (g)	Calcium (mg)	Phosphorus (mg)	Iron (mg)
1.	Agathi	93	73	8	1	3	2	12	1130	80	4
2.	Amaranths	26	90	3	1	3	1	6	397	83	3
3.	Ash gourd	10	96	0	0	0	2	1	30	20	1
4.	Beet root	43	88	2	0	1	9	1	18	55	1
5.	Betel leaves	44	85	3	1	2	2	6	230	40	11
6.	Bitter gourd	25	92	2	0	1	4	1	20	70	1
7.	Bottle gourd	12	96	0	0	0	2	1	20	10	0
8.	Brinjal	24	93	1	9	9	4	1	18	47	0
9.	Broad beans	48	85	4	0	1	7	2	50	64	1
10.	Broad bean leaves	71	78	6	0	1	4	11	111	149	0
11.	Brussels sprouts	52	85	5	0	3	2	13	340	110	9
12.	Cabbage	27	92	2	0	1	1	5	39	44	1
13.	Carrot	48	86	1	0	1	11	1	80	530	1
14.	Carrot leaves	77	77	5	0	3	2	13	340	110	9
15.	Cauliflower	30	91	3	0	1	4	1	33	57	1
16.	Cauliflower leaves	66	80	6	1	3	2	8	626	107	40

Contd.

17.	Celery leaves	37	88	6	1	2	1	2	230	140	6
18.	Celery stalk	18	93	1	0	1	1	3	30	38	5
19.	Cho-Cho	27	92	1	0	0	6	1	140	30	1
20.	Cluster beans	16	81	3	0	1	11	3	130	57	1
21.	Colocasia	97	73	3	0	1	18	0	16	31	0
22.	Colocasa leaves	56	83	4	1	2	3	7	227	82	10
23.	Colocasia stem	18	94	0	0	1	4	1	60	20	0
24.	Cowpea pod	48	85	3	0	1	8	2	72	59	2
25.	Cucumber	13	96	0	0	0	2	0	10	25	1
26.	Coriander leaves	44	86	3	1	2	1	6	184	71	1
27.	Cow pea leaves	38	89	3	1	2	1	4	290	58	20
28.	Curry leaves	108	63	6	1	4	6	19	830	57	1
29.	Drumstick	26	87	3	0	2	4	5	30	110	0
30.	Drumstick flowers	50	86	4	1	1	7	1	51	90	0
31.	Drumstick leaves	92	76	7	2	2	1	12	440	70	1
32.	Fenugreek leaves	49	86	4	1	1	1	6	395	51	2
33.	French beans	26	92	2	0	1	5	2	50	28	1
34.	Karonda fresh	42	91	1	3	0	3	2	21	28	0
35.	Karonda dry	364	18	2	110	3	67	0	160	60	39
36.	Knol-Khol	21	93	1	0	1	4	2	20	35	2
37.	Knol-khol greens	43	87	3	0	1	2	6	667	99	17
38.	Lady finger	31	90	2	1	1	14	3	67	25	0

Contd.

39.	Leeks	77	79	2	0	1	17	1	50	70	2
40.	Lettuce	21	93	2	0	1	0	2	50	28	2
41.	Lotus stem, dry	234	10	4	1	9	51	25	405	128	61
42.	Mango ginger	53	85	1	1	1	10	1	25	90	3
43.	Mango, green	44	88	1	0	0	10	1	10	19	0
44.	Mint	48	85	5	1	2	2	6	200	62	16
45.	Mustard leaves	34	90	4	1	2	1	3	155	26	16
46.	Onion big	50	87	1	0	0	11	1	47	50	1
47.	Onion small	59	84	2	0	1	13	1	40	60	1
48.	Onion stalks	41	88	1	0	1	9	2	50	50	7
49.	Parsley	87	75	6	1	3	2	13	390	175	18
50.	Parsnip	101	72	1	0	1	23	2	50	40	0
51.	Pumpkin flowers	39	89	2	1	1	6	1	120	60	0
52.	Pumpkin fruit	25	93	1	0	1	5	1	10	30	0
53.	Pumpkin leaves	57	82	5	1	3	2	8	392	112	-
54	Radish leaves	28	91	4	0	2	1	2	265	59	0
55.	Radish pink	32	91	1	0	1	7	1	50	40	0
56.	Radish white	17	94	1	0	1	3	1	35	22	0
57.	Ridge gourd	17	95	1	0	0	3	0	18	26	2
58.	Safflower leaves	33	91	2	1	1	-	4	185	35	6
59.	Snake gourd	18	94	1	0	1	3	1	26	20	22
60.	Sweet potato	120	68	1	0	1	28	1	46	50	0

Contd.

61.	Sword beans	44	87	3	0	1	8	2	60	40	2
62.	Spinach	26	92	2	1	2	1	3	73	21	1
63.	Spinach stalks	20	93	1	0	2	-	4	90	20	2
64.	Tamarind leaves	115	70	6	2	1	2	18	101	140	0
65.	Tapioca	157	59	1	0	1	38	1	50	40	1
66.	Tomato, green	23	93	2	0	1	4	1	20	36	1
67.	Turnip	29	92	0	0	1	6	1	30	40	0
68.	Turnip greens	67	82	4	1	2	1	9	710	60	28
69.	Water chestnut, fresh	115	70	5	0	1	23	1	20	150	2
70.	Water chestnut, dry	330	14	13	1	3	69	0	70	440	2
71.	Yam, elephant	79	79	1	0	1	18	1	50	34	1

Source :

Gopalan. C, Rama Sastri B.V. and Balasubramanian, S.C., 2004, Nutritive Value of Indian Foods, National Institute of Nutrition, ICMR, Hyderabad.

Appendix-VI

Classification of Spices based on Morphological Characters

Morphology of the useful part	Example of the Spices
Berry	Black pepper, All spice, Juniper
Seeds	Aniseed, Caraway, Celery, Coriander, Fenugreek
Fruit	Cardamom, Capsicum
Aril	Mace
Kernal	Nutmeg
Rhizome	Ginger, Turmeric
Leaf	Basil, Bay leaf, Tejpath, Curryleaf
Bark	Cinnamon, Cassia
Bulb	Garlic, Onion
Rind	Garcinia, Kokam
Stigma	Saffron
Pod	Vanilla
Flower bud	Clove
Oleogum resin	Asafoetida
Root Fruit pulp	Caper
	Tamarind

Appendix-VII

Various Structures of Flowers

Flower	Structure
Poinsettia	Bract
Cymbidium Orchid	Modifications in labellum's petal as well as fusions of Stamens and Pistils
Rose	Complex/Single whorl of Petals
Eucalypt	Prominent feature stamens
Anthurium	Spadix Plus Spathe
Grevillea	Raceme
Paper Daisy	Head
Belladonna Lily	Umbel
Bottlebrush	Spike
Statice	Panicle
Kangaroo Paw	Cyme
Hybrid rose	Solitary
Spray Carnation	Corymb

Appendix-VIII

Classification of fruits according to their respiratory behavior during ripening

Climacteric Fruits	Non-Climacteric Fruits
Apple, Apricot, Annona , Avocado, Bael, Banana, Blue berry, Cranberry, Durian, Fig, Guava, Jack Fruit , Kiwi fruit Mango, Mangosteen, Muskmelon, Papaya, Passion Fruit, Peach, Pear, Plum, Rasp berry, Sapota, Tomato, Watermelon etc.	Aonla, Ber, Carambola, Cherry, Citrus, Cucumber, Grape, Jamun, Litchi, Loquat, Nut, Olive, Pineapple, Phalsa, Pomegranate, Pumpkin, Rambutan, Strawberry.

Appendix-IX

Classification of fruits and vegetables based on Respiration rate

Respiration Rate	Rep. Rhythm . mg.CO_2/kg.h	Fruit /Vegetable
Very Low	<5 mg	Nut, dried fruits and vegetable, potato garlic and onion
Low	5 - 10 mg	Apple, citrus, grape, sugar beet, turnip, sweet potato, watermelon and cucumber.
Moderate	10 - 20 mg	Mango, Banana, peach, pear, fig cabbage, carrot, tomato, and carrot.
High	20 - 40 mg	Avocado, cauliflower, lettuce, strawberry and raddish.
Very High	40 - 60 mg	Artichoke, cut flowers, parsley, sweet corn, leek, brusselsprout and snap melon.
Extremely High	>60 mg	Spinach, asparagus, green pea, mushroom and broccoli.

Appendix-X

Heat processing requirements for different products based on different acidity and pH

Acidity class of food	pH	Food item	Processing requirements
Low acid	5.0 – 7.0	Peas, carrots, beets, potatoes, asparagus, tomato soup	High temperature processing 116-121°C (240-250°F)
Medium acid	3.7 - 4.5	Tomatoes, pears, apricots, peaches	-do-
Acid	3.0 - 3.7	Sauerkraut, pineapple, apple	Boiling water processing 100°C (212°F)
High acid	3.0 – 2.3	Pickles, lime juice	-do-

Appendix-XI

Fruits and Vegetables with Aroma compounds

Fruit and Vegetable	Compound
Apple - Ripe	Ethyl 2-methylbutyrate
- Unripe	Hexenal, 2-hexenal
Banana - Green	2-Hexenal
Ripe	Eugenol
Overripe	Isopentanol
Blueberry	Trans-2-hexanal
Grape	Methyl anthranilate, Ethyl acetate
Grapefruit	Nootakatone
Lemon	Citral
Lime	Limonene
Orange	Valencene
Pear	Decadienoate esters
Raspberry	1-(π-Hydroxyphenyl)-3-butanone
Cucumber	2,6-Nonadienal
Cabbage - Raw	Allyl isothiocyanate
- Cooked	Dimethyl disulphide
Celery	Dihydrophthalides
Garlic	Diallyl disulphide
Mushroom	1-Octen-3-ol, lenthionine
Onoin	allyl propyl disulfide
Parsley	Monoterpene hydrocarbons
Potato	2-Methoxy-3-ethyl pyrazine, 2.5-dimethyl pyrazine
Radish	4-Methylthio-tans-3-butenyl isothiocyanate
Soyabean	Ethyl vinyl ketone

Appendix-XII

Classification of fruits and vegetables based on perishable nature

Perishability Index	Potential Life (Weeks)	Fruits	Vegetables
Very high	< 2 weeks	Blackberry, strawberry etc.	Broccoli, cauliflower,
High	2 - 4 weeks	Avocado, pineapple, etc.	Celery, tomato
Moderate	4 - 8 weeks	Lemon, watermelon, mango, etc.	Potato
Low	8 - 16 weeks	Apple, pear etc.	Onion, garlic,
Very low	> 16 weeks	Nuts and dried fruits	

Appendix-XIII

Government of India approvals for Irradiation dose requirements for fruits and vegetables

Commodity	Irradiation dose (kGY)		
	Minimum	Maximum	Purpose
Onion	0.03	0.09	Sprout inhibition
Potatoes	0.05	0.15	Sprout inhibition
Raisins	0.25	0.75	Insect disinfestations
Dried fruits	0.25	0.75	Insect disinfestations
Mango quarantine treatment	0.25	0.75	Shelf life extension and
Ginger	0.03	0.15	Sprout inhibition
Garlic	0.03	0.15	Sprout inhibition
Small onions (shallots)	0.03	0.15	Sprout inhibition

Appendix-XIV

Chilling injury symptoms of some Fruits and Vegetables

Commodity	Approx. temp. (°C) below which chilling injury starts	Symptoms
Banana green	12	Browning of skin, smoky appearance, browning of the pulp, failure to ripen at RT.
Papaya	12	Discolouration of the skin, pitting, water soaking of flesh, failure to ripen at RT, off-flavour development, poor carotene development.
Mango	10	Discolouration and browning of the skin, pitting of the skin, failure to ripen at RT, poor carotene development in the flesh.
Pineapple	15	Brown or dull shell colour, water soaked flesh, wilting of crown, browning of tissue near the core and failure to develop full flavour at RT.
Lime	9	Pitting and browning of the peel, darkening of the oil glands in the skin, and browning of the flesh.
Capsicum	7	Pitting of the skin, soggy spots in the skin.
Tomato green	10	Failure to develop red colour at RT, pitting of the skin, failure to ripen at RT.

Appendix-XV

Derivatives of some vegetables from plant tissue

Fruit	Plant Tissue
Artichoke	Flower Bud
Asparagus	Stem Sprout
Beetroot	Swollen Hypocotyls
Broccoli	Swollen inflorescence
Brussels sprout	Auxiliary Bud
Carrot	Swollen Root
Celery	Petiole
Corn	Seed
Onion	Bulb (Underground Bud)
Leek	Swollen Leaf Base
Lettuce	Main Bud
Potato	Stem Tuber
Pumpkin	Fruit
Spinach	Leaf Blade
Sweet Potato	Swollen root tuber

FSC
www.fsc.org
MIX
Papier | Fördert
gute Waldnutzung
FSC® C083411

Zeitfracht Medien GmbH
Ferdinand-Jühlke-Straße 7
99095 Erfurt, Deutschland
produktsicherheit@kolibri360.de